THE COMMON AGRICULTURAL

POLICY AND ORGANIC FARMING

THE COMMON AGRICULTURAL

POLICY AND ORGANIC

FARMING

K. Lynggaard

www.cabi.org

CABI Publishing is a division of CAB International

CABI Publishing
CAB International
Wallingford
Oxfordshire OX10 8DE
UK

Tel: +44 (0)1491 832111
Fax: +44 (0)1491 833508
E-mail: cabi@cabi.org
Website: www.cabi.org

CABI North American Office
875 Massachusetts Avenue
7th Floor
Cambridge, MA 02139
USA

Tel: +1 617 395 4056
Fax: +1 617 354 6875
E-mail: cabi-nao@cabi.org

A catalogue record for this book is available from the British Library, London, UK.

A catalogue record for this book is available from the Library of Congress, Washington, DC.

ISBN   1 84593 114 9
         978 1 84593 114 8

Printed and bound in the UK from copy supplied by Athenaeum Press Ltd.

Contents

List of Tables

List of Abbreviations

Agri. Commissioner	Commissioner for Agriculture
AT	Austria
BOF	British Organic Farmers
BSE	Bovine Spongiform Encephalopathy
CAP	Common Agricultural Policy
CEPFAR	European Training and Development Centre for Farming and Rural Life
COGECA	General Committee for Agricultural Cooperation in the European Union
COPA	Confederation of Professional Agricultural Organisations
DE	Germany
DG Agri.	Directorate-General for Agriculture
DG Env.	Directorate-General for the Environment
DG	Directorate-General (of the European Commission)
DK	Denmark
EC	European Community
EE	Estonia
EEB	European Environmental Bureau
EL	Greece
Env. Commissioner	Commissioner for the Environment
EP Com. Agri.	EP Committee on Agriculture, Fisheries and Food
EP Com. Env.	EP Committee on the Environment, Public Health and Consumer Protection
EP Com. Reg.	EP Committee on Regional Policy and Regional Planning
EP	European Parliament

ES	Spain
EU	European Union
FI	Finland
FR	France
GATT	General Agreement on Tariffs and Trade
GMO	Genetically Modified Organism
IE	Ireland
IFOAM	International Federation of Organic Agriculture Movements
LT	Lithuania
MAFF	Ministry of Agriculture, Fisheries and Food (UK)
MEP	Member of European Parliament
NFU	National Farmers Union (UK)
NL	The Netherlands
NO	Norway
QMV	Qualified Majority Vote
SE	Sweden
SEA	Single European Act
SEM	Single European Market
TEU	Treaty on European Union
UK Agri. Com.	House of Commons Agriculture Committee (UK)
UK	United Kingdom
UKROFS	UK Register of Organic Food Standards
US	United States
WTO	World Trade Organisation
WWF	World Wide Fund for Nature

Preface

The theoretical concern of this book is the question of how institutional change may be captured and conceptualised: in particular, how we may capture and conceptualise the type of institutional change which is ideational in nature. Empirically, the study deals with institutional change within the Common Agricultural Policy of the European Union. In this regard, a specific focus is on the institutional construction of a policy field concerned with organic farming within the auspices of Common Agricultural Policy from the late 1970s/early 1980s to 2005.

My work on the book has benefited greatly from advice and inspiration from a number of people. I would like, in particular, to thank Associate Professor Johannes Michelsen (University of Southern Denmark), Professor Wyn Grant (University of Warwick), and Professor O.K. Pedersen (University of Copenhagen). I would also like to thank Carolyn Foster, Susanne Padel and Nic Lampkin (University of Wales) for making comprehensive archival material available to me and for their hospitality during my visit at the Organic Centre Wales. Finally, I would like to thank a number of people, whom at various stages have been very helpful with proofreading, translation and layout: Gitte M. Kaastrup (English), Sheila Kirby (English), Søren Riis (German), Kirsten Verkooijen (Dutch) and Janni Villadsen (Layout).

Kennet Lynggaard
Roskilde, November 2005

1

The Common Agricultural Policy and Institutional Change

This book has a general concern with the conditions for, and dynamics of, institutional and ideational changes and a specific concern with changes in the Common Agricultural Policy (CAP) of the European Union (EU). Starting with the specific concern, the CAP is often thought of as a political field characterised by a very high degree of stability in terms of polity, politics and policy. This perception is evident in references to the CAP which has been described variously as: being 'extremely path dependent' and governed by a 'club-like' Agriculture Council (Peterson and Bomberg 1999, p.135; Swinbank 1989, p.304), as governed by an 'iron triangle' of farmers' organisations, Commission officials and Ministers of Agriculture (Hix 1999, p.253), as having found its 'policy equilibrium' (Peters 1996, p.63) or, in general, as 'a prime example of multiple forms of institutional resistance to change' (Banchoff 2002, p.6). The gist of the matter is that the CAP is often considered to be very much set as to the types of agents participating in the formulation of the policy, the political processes guiding the field and there also seems to exist clear boundaries on which issues are to be included and excluded from consideration. On the other hand, recent research into the development of the CAP has increasingly pointed to changes in its institutional set up and to changes which may be regarded as ideational. In regard to changes in the institutional set up, it has been observed that agents other than those traditionally considered to govern the CAP have become involved in CAP decision-making including representatives of industrial and environmental interests, and groupings within the European Parliament (EP) (Nedergaard *et al.* 1993, chapter 5; Hennis 2001; Roederer-Rynning 2003). Changes have also been observed in relation to the processes guiding agricultural decision-making. For instance, it appears that the CAP has opened up to the influence of international trade negotiations, which are often signalled out as being significant for recent changes within the CAP (Patterson 1997; Coleman and Tangermann 1999; Sheingate 2000; Hennis 2001; Coleman *et al.* 2004). It also seems to have become increasingly common to refer agricultural matters – when the level of conflict in the Agriculture Council is high – to the Council of the European Union (Kay 1998, chapter 3; Ackrill 2000, chapter 5).

In regard to ideational change, it appears that new types of problems have entered the CAP agenda and untraditional solutions have been proposed, some of which have been adopted as community policy. In general terms, it has been suggested that agricultural problems have been redefined and turned into societal problems (Nedergaard *et al.* 1993; Sheingate 2000). More specifically, policies have been adopted in regard to, for instance, the environment (Gardner 1996, chapter 6; Grant 1997, pp.200–213; Lowe and Whitby 1997; Greer 2005), food safety issues and rural development (Skogstad 2001; Roederer-Rynning 2003; Greer 2005). Although unusually bold, initiatives to reform the CAP in the early 1990s have been characterised as a shift that 'represented a fundamental change in the philosophy underlying the CAP' (Patterson 1997, p. 135; see also Paarlberg 1997).

Altogether, on the one hand, the CAP is, seemingly, a highly stable policy field. On the other hand, the body of literature concerned with CAP reform has increasingly observed that changes have occurred within this field: changes which – as will be expanded on below – may be regarded as institutional and ideational in nature. Yet, while a number of conceptualisations of continuity have already been offered by the literature concerning CAP reform, there still exist a series of theoretical and methodological challenges as to developing coherent and sensitive concepts of change and, thus, account for the empirical observations made about developments within the CAP. This paradox is not only of analytical interest but it is also necessary to address in order to enhance our understanding of institutional and ideational change within the CAP and, ultimately, to make plans for and implement change.

Continuity, Change and the Common Agricultural Policy

Six major attempts to reform the CAP have been made since the policy was first launched in 1958. The first was the 1968 'Mansholt Plan'. This was followed in the 1980s by the Milk Quota Reform (1984) and the Budget Stabilisers Reform (1988) and, in the 1990s, another two attempts were made: the MacSharry Reform (1992) and Agenda 2000 (1999) (see e.g. Kay 1998, chapter 3; Ackrill 2000, chapter 5). Most recently, an attempt to reform the CAP was made in 2003 (Commission 2003). Against this background, a series of political science studies have addressed recurrent attempts to reform the CAP through various theoretical optics. The following will enlarge on the paradox outlined above by means of a review of approaches to the study of continuity and change and suggest a number challenges facing the literature concerning CAP reform through a critical discussion of the explanations and conceptualisations of change offered by this body of literature. It should be noted that the main focus of interest of the body of literature reviewed, whilst not to the exclusion of the discussion of other reforms, is that of the 1992 CAP reform, which is generally regarded as the greatest departure from the status quo.

The first two sections (Network, Ideas and Policy Frames; Venues and Issues) will consider approaches to the study of CAP reforms which, to varying degrees, have a concern with the role of ideas in the development of the CAP. Hereafter, there follows a section on approaches that emphasise a relationship between the various levels of negotiation (Interlinked Developments at Various Levels of Negotiation), a section on approaches which have their point of departure in rational choice theory (Public-Choice and Principal-Agent Relationships), a section considering approaches to the study of CAP reform, which are eclectic in the sense that, in order to fully understand various CAP reforms, it is proposed that different types of theoretical models may be needed (Eclectic Approaches) and, finally, a section, which sums up the explanations of continuity and change offered by the literature concerned with CAP reform (Summing up on Explanations of Continuity and Change). The chapter will be concluded by a critical discussion, which throws down five challenges to the methodologies and concepts offered by the body of literature concerned with CAP reform in an endeavour to capture ideational change (Five Challenges to the Study of Ideational Change) and, finally, an outline of the book.

Networks, Ideas and Policy Frames

The CAP is often considered largely resistant to outside influence, which may seek to effect a change in the CAP, due to existence of cohesive policy networks within the CAP (Grant 1993; Daugbjerg 1998, 1999; Pappi and Henning 1999). The main characteristic of cohesive policy networks like those claimed to be found within the CAP is the existence of a pronounced consensus among network members about a basic policy paradigm. Within such networks it is very unlikely that advocates of fundamental changes will be successful and a general reluctance exists to set the policy in motion and take the risk of changes that may have unforeseen and unwanted consequences. Daugbjerg (1999) locates these meso-level policy preserving networks within a set of institutional structures on the EU macro level, which also disfavours change. The fragmented nature of the political system of the EU and the absence of one centre of authority place severe restrictions on the potential of generating the power necessary to carry through reforms and increase the number of power centres that may oppose reform attempts. Against such a background, the 1992 reform is described as the most far-reaching one. Yet the changes cannot, according to Daugbjerg (1999), be labelled as being radical or fundamental since the reform did not contest the extensive use of subsidies. The main points, made by Daugbjerg are that attempts to reform the CAP have had only limited success as this policy field is very resistant to change and, regarding the extent to which changes have taken place, these can be viewed as incremental. The reason for this path of development is the nature of the meso-level policy networks imbedded in the CAP as well as the macro institutional framework within which the CAP develops.

Whereas both studies of policy networks and policy paradigms within the CAP operate within an institutional analytical framework, approaches emphasising policy paradigms or policy frames (Lenschow and Zito 1998; Skogstad 1998) rather than policy networks tend to replace policy networks with ideas as the prime independent variable in explaining continuity and change. Accordingly, in a comparison of attempts to reform EU and US agricultural policy Skogstad (1998) points to the importance of ideas in shaping policy outcomes. It is implied that the likeliness of change will increase, (i) when the ideas informing a policy field have a low degree of institutionalisation; (ii) if these ideas are in opposition to wider societal rationalities; and (iii) when the policies and policy instruments put into force with reference to these ideas have a disintegrating effect on the regulated sector (Skogstad 1998, p.465; see also Coleman *et al.* 1997). The CAP is, however, characterised by the opposite and the reasons for the continuity of the CAP are ideational. The idea that the agricultural sector – for various reasons – should be supported by the EU is constitutionally imbedded in the Treaty of Rome.[1] This idea is supported by the strong and highly institutionalised relationships that exist among farmers' organisations, agriculture officials, and politicians on both national and supranational levels (Skogstad 1998, p.479). Furthermore, the basic aim of supporting a common market for agricultural products within the EU, the inclusion of policies under the auspices of the CAP regulating most agricultural products and the high level of organisation among farmers in the EU Member States has a unifying effect on the agricultural sector (Skogstad 1998, p.480). Finally, it is claimed that the continuity of the CAP is enhanced by a broad societal acceptance – in particular in France and Germany – of the idea that agriculture is a matter of state interest and should be supported as such. Agriculture is thought to be an exceptional sector operating under unpredictable

---

[1] The Treaty of Rome established the European Economic Community in 1957.

conditions (weather and unstable markets) and as being a matter of national interests. Hence, it is widely accepted that there are legitimate reasons for supporting this sector (Skogstad 1998, p.481). Altogther, Skogstad claims that while successful and fundamental reforms, directed at the liberalisation of agricultural markets, have been carried through in the US, the CAP did not experience equivalent changes during the 1990s and, although the MacSharry reform introduced 'new, market-liberalizing policy instruments, the underlying goals of the CAP remained intact' (Skogstad 1998, p.463).

Lenschow and Zito (1998) see the 1992 MacSharry reform as a consequence of events external to the CAP. The theoretical framework applied emphasise a typology of environmental policy frames in order to capture dissimilar ways in which agents 'make sense' of and act in very complex political contexts. On one hand, the institutional characteristics of a political system will shape the flow of ideas and, on the other hand, policy frames will be rooted in the institutional structures. The institutional structures, which essentially shape policy outcomes, are seen as having an organisational, procedural and normative dimension.

From this perspective, the 1992 Macsharry reform is seen as a consequence of the establishment of the Single European Market (SEM), the enlargement of the EU and the pressure generated by international negotiations on trade liberalisation within the General Agreement on Tariffs and Trade (GATT; from 1994 the World Trade Organisation; WTO). More specifically, the SEM project and the Cohesion Policy that followed paved the way for criticism related to the financial burden of the CAP, the distortion of financial support in favour of efficient northern farmers, and the adverse environmental effects of the CAP. Enlargement to the south and potentially to the east have also highlighted the CAP expenditures and having had a fragmenting effect on the farm lobby, which is increasingly pursuing dissimilar interests. Finally, support to environmentally friendly farming has been introduced but it is claimed that this is mainly to legitimise continuing support for agriculture in the context of the GATT as well as in the broader public arena. By and large, Lenschow and Zito (1998) claim that the CAP has been only very marginally moved towards a policy more informed by environmental concerns and suggests that the 'thick institutional structure' of the CAP is the reason why real institutional changes in favour of such a development have not appeared. Organisationally, links have been established with the Directorate-General (DG) for the Environment adding to the complexity of the CAP but still leaving the policy highly sectorial. The new procedural structures on environmental policymaking introduced by the Single European Act (SEA) in 1987 and the Treaty on European Union (TEU) in 1992 lacked, according to Lenschow and Zito, clear definitions disfavouring their usage and, hence, worked against the intended change. Along the normative dimension it is claimed that the 'politically influential farming sector is still some distance away from accepting its responsibility for environmental degradation and the need to integrate environmental considerations into production processes' (Lenschow and Zito 1998, p.438).

Venues and Issues

As it appears, both studies of the CAP with their point of departure in analytical frameworks investigating the independent effect of policy networks and policy paradigms or policy frames on developments within the CAP supply explanations of continuity within this policy field. Sheingate (2000), however, makes a break away from addressing the question of why the CAP has shown itself to be highly resistant to reform attempts and, rather, pursues the question of how changes in terms of agricultural retrenchment have come about within the CAP. That is, basically, continuity is replaced by change as the dependent variable. Against that background, Sheingate – using Baumgartner and Jones (1993) – proposes that the concepts of issue definition and venue change may improve our understanding of change. It is claimed that if a political issue is redefined then new aspects of the issue will surface. Such a redefinition of an issue may destabilise the majority that backed the original issue definition and, in essence, the destabilisation of this majority may bring about change. The redefinition of an issue may also pave the way for change in the venue dealing with a particular issue. Where the jurisdiction over a particular policy is not clear-cut, the venue, which eventually becomes the forum within which the policy is dealt with, may very well be significant in terms of whether changes in the policy may take place or not. Skilful policy entrepreneurs seeking policy changes play an important role in this regard. They will act strategically and seek to redefine the issue at hand in order to destabilise the majority that prefers the status quo and/or will direct the issue towards a venue, which is considered more in favour of change.

Applying these concepts Sheingate finds that the CAP – compared with US agricultural policy – is more reluctant to change since there exist clear jurisdictional lines in the area that hinder venue change. It is uncontested that agricultural related issues are dealt with by the DG for Agriculture, the Special Committee for Agriculture, and the Council of Agriculture Ministers. On the domestic level – particularly in France and Germany – agricultural policy takes place in neo-corporatist environments made up by farmers' interest organisations and the national bureaucracies of Agriculture Ministries. These are general considerations relating to the continuity of the CAP. However, the 1992 reform is the exception from the rule and exemplifies how turning to the venue of the GATT made a difference. This, Sheingate argues, was amplified by the mere threat of additional venue change from Ministers of Finance, Industry and Foreign Affairs and from Heads of Government who would take over control of the negotiations within the GATT if the traditional agents in charge of agricultural policymaking did not find a solution (Sheingate 2000, p.355). Finally, the reason why venue change (and threats of such) was a real concern, can be found in the successful redefinition of agricultural issues which occurred in the time leading up to the 1992 CAP reform. Sheingate argues that agricultural issues had largely been redefined and it had become an increasingly widespread perception that the central problems or 'negative externalities' of the existing CAP were the ever-increasing expenditures as well as surplus production. Hence, the main explanation of change offered by the theoretical framework emphasising the interrelationship between issues and venues is related to perceptual or what may be seen as ideational changes, which have created new expectations of what kind of problems agricultural policies ought to deal with.

Interlinked Developments at Various Levels of Negotiation

Continuity and change within the CAP is often explained by unlike but interlinked 'games' on the international, the EU and sometimes the domestic (Member State) levels (Paarlberg 1997; Patterson 1997; Coleman and Tangermann 1999; albeit less explicit see also Landau 1998). While Paarlberg (1997) only ascribes very little causal effect to the 'Uruguay Round' of the GATT on changes within the CAP,[2] Coleman and Tangermann (1999) suggest that the 1992 CAP reform was a product of the autonomous but linked negotiations as they took place in the context of the CAP and the GATT respectively. On one hand, the games or negotiations in each forum are autonomous as they are imbedded in a particular set of institutional rules and involve different sets of agents each of which are concerned with specific policy problems. On the other hand, the negotiations are linked, since some agents are involved in both games, and the outcome of each game will depend on that of the other (Coleman and Tangermann 1999, p.388). Through an analysis of the negotiations leading to the CAP reform in 1992 and those leading to the finalisation of the Uruguay Round in 1994, Coleman and Tangermann found that if the CAP had not been influenced by the GATT negotiations, it is unlikely agreement would have been reached to move from price support, towards a greater emphasis on direct farmer income support, and the introduction of a compulsory set-aside scheme. In spite of an institutionally path dependent CAP, and the presence of forceful agricultural policy networks that acted against radical changes, the CAP was reformed due to the entrepreneurial role played by the Commission and, essentially, 'shaped by the interpretations of what policies might be successfully enshrined in a GATT agreement' (Coleman and Tangermann 1999, p.401). It is a simple point, the possible outcome of the CAP reform would have been different, had it not been for the involvement of the European Commission and the parallel struggle between the US and the EU in

---

[2] Paarlberg (1997) suggests that the parallel negotiations in the context of the GATT expedited the 1992 CAP reform process but, essentially, changes within the CAP would have come about regardless.

the context of the GATT. Opposed to Coleman and Tangermann's two-level analysis, Patterson (1997) suggests that the CAP reforms of 1988 and 1992 should be analysed as three-level games where negotiations take place domestically, at the EU level and at the international level. The basic line of argument is that:

[b]ecause domestic coalitions affect the passage of Community agricultural policy and Community agricultural policy affects world markets, and because world market conditions affect domestic coalitions and Community agricultural policy, the policy shift that occurred in 1992 [opposed to the 1988 reform] can be explained only by changes that occurred simultaneously at several levels of the game (Patterson 1997, p.142).

The two central concepts, used to capture the dynamic of change, are 'synergistic linkages' and 'reverberation' (cf. Putnam 1988). The concept of reverberation seeks to capture the dynamic of how 'strategies and outcomes at different levels of the game [are] simultaneously affecting one another' (Patterson 1997, p.142) and 'implies that international pressure expands the domestic win set and facilitates agreements' (Patterson 1997, p.151). Synergistic linkages refer to the dynamic, which may appear when issues are coupled at the international or Community level and, essentially, alter the types of outcomes which are considered viable by domestic constituencies (Patterson 1997, p.151). That is, whereas reverberation implies a change in domestic preferences, synergistic linkages create new policy options, which would not have been possible in a pure domestic context.

In a comparison between the 1988 and 1992 reform Patterson claims the 1988 reform as an incremental change of minor importance. However, the 1992 reform is unusually boldly characterised as a shift that 'represented a fundamental change in the philosophy underlying the CAP' (Patterson 1997, p.135). According to Patterson, the main reasons for this were dissimilar but interlinked developments on the international level, the Community level and the domestic level. Internationally, by 1992 it had become clear that world agricultural trade had not been improved by the 1988 CAP reform and that the GATT negotiations would collapse altogether if a compromise was not made on agricultural issues. On the Community level the budget pressure of the CAP had not been solved by the 1988 reform and at the same time Community activities were expanding as a consequence of the implementation of the SEA and the adoption of the TEU.

On the domestic level, German reunification created great pressures on the German budget, diversified the farm lobby and, at the same time, interests other than those of the agricultural sector, in particular labour interest, were increasingly making demands on the government. In France, the government was increasingly aware of an imminent Community budget crisis if agricultural production and expenditures continued to rise and it was considered better to back a cut in support prices than production quotas, which were feared would lead to a reduction in the share of agricultural markets of the French farmers both domestically and internationally. The UK backed a CAP reform as before in 1988 but, importantly, at this point in time, the German government had a broad concern which allowed them to support, according to Patterson, radical CAP reform. Altogether, Patterson suggests that the 1992 reform was enabled by the linking of various levels of negotiation, a significant increase in the real and perceived costs of not reforming the CAP, and a diversification of interest group pressures: in particular in Germany. Key players – in particular the Commissioner for Agriculture, Ray MacSharry, and the German Chancellor, Helmut Kohl – were critical in establishing links between the various levels of negotiations.

Public-Choice and Principal-Agent Relationships

It is also possible to identify a group of studies within the CAP reform literature which explicitly ascribe to some sort of rational choice or rational choice institutional school of thought. Along these lines, Nedergaard *et al.* (1993) approach the study of the CAP through a public-choice perspective. The point of departure here is that an important motive for individual behaviour is a concern about economic gains. It is assumed that collectives such as various administrative units, interest organisations and governments can be studied as if they are seeking to maximise their material utility insofar as these collectives are inhabited by individuals with a coherent set of interests. Institutional constraints imposed by public decision-making procedures on individual behaviour are claimed to be central as to how the preferences of individuals are turned into public policy (Nedergaard *et al.* 1993, pp.76–77). The types of explanations offered by this theoretical perspective are related to 'government failures'. For instance, the privileged access of farmers' interest organisations to political institutions have produced a protectionist policy with the resultant costs carried by taxpayers, who constitute a group that is less well organised and inclined to invest in the decision-making process in order to keep the expenditures of the CAP from rising (Nedergaard *et al.* 1993, pp.97–99). Moreover, since the costs of the CAP budget are carried collectively, Member States tend to maximise national economic interest by claiming the highest possible return through the expansion of national agricultural production (Nedergaard *et al.* 1993, pp.126–132). Changes, which are considered to have taken place since the early 1990s, are, in essence, claimed to be a product of shifts in 'supply and demand behaviour on the political markets' (Nedergaard *et al.* 1993, p.147). Importantly, the continuously increasing expenditures made to the agricultural sector, problems of surplus production, and the adverse environmental effects of the CAP, have turned agricultural policy into a societal problem. In addition, the Commission as well as the EP have been given increased powers in the EU decision-making process by the treaty revisions in 1987 and 1992, which is a development considered to favour changes in the CAP. International negotiations within the GATT have put pressure on the CAP itself, but have also given rise to other stakeholders such as industrial interests concerned with avoiding international trade conflicts caused by EU agricultural protectionism. The implementation of the SEM has likewise given rise to concerns other than mere agricultural ones in the EU and the approaching eastern enlargement is thought to necessitate further CAP reform (Nedergaard *et al.* 1993, chapter 5).

The 1992 CAP reform is broadly agreed to constitute the greatest departure from the status quo and this is often attributed to international pressure for change during the Uruguay Round negotiation within the GATT, which took place while the CAP reform was in the preparatory phase. Applying a principal-agent model, Pollack (2001) has studied the relationship between Member State representatives (the principals) and the Commission (the agent) during the Uruguay Round negotiation in order to evaluate the role of the Commission and the degree of latitude it has in carrying through a CAP reform. Although a series of administrative oversight procedures have been established by the Member States to ensure that the supranational agents behave as intended by their creators (Pollack 1997, pp.108–121), the agents may still have some latitude to pursue their own preferences even when these are not in accordance with those of their principals. Thus, agency loss may arise in the form of either 'shirking' or 'slippage'. 'Shirking' appears when the agents seek to optimise their own preferences at the expense of the preferences of their principals, whereas 'slippage' refers to the unintended consequences of an incentive structure of a delegation that leads agents to behave contrary to what was intended by the principals when they decided on a particular principal-agent relationship (Pollack 1997, p.108; 2001, p.220).

Pollack finds that, first, the latitude of the Commission to carry through a radical reform of the CAP was severely constrained due to strong oppositional preferences among Member States – the opposition from France being particularly intense. Second, the norm of consensus decision-making within the Agriculture Council – despite the Qualified Majority Vote (QMV) being the legally required decision-making procedure – in practise attributed each Member State with a veto and thus significantly reduced the possibility for reform. Third, whereas an agent is usually considered to possess an informational advantage in relation to, for example, the policy positions of the involved parties that may be used strategically in 'shirking' vis-à-vis the preference of the principals, the Commission did not uphold this advantage in the case considered. Rather, Member State representatives were able to monitor and define the mandate of the Commission during the GATT negotiations within the Art. 113 Committee[3] and thus left little room for the Commission to pursue its own agenda. Fourth, the Commission could not claim to possess the support of any particular transnational coalition – such as multinational business associations or transnational interests groupings – in its endeavour to reform the CAP, which could otherwise had been exploited as a resource during negotiations (Pollack 2001, p.221, 246–247).

Whereas the mentioned institutional constraints all disfavour the Commission's independence to pursue its own preferences, Pollack also points out that 'the Commission was both purposeful and successful in harnessing external US pressures and internal budgetary pressures to produce and steer through the Council a reform of the CAP more rapid and more far-reaching than the Council would likely have adopted in the absence of Commission entrepreneurship' (Pollack 2001, p.246). Hence, according to Pollack, although the degree of latitude of the Commission was limited during the parallel negotiations on trade liberalisation in the GATT and on agricultural reform within the EU, the Commission was still the agent of change while Member States defended the status quo. There existed clear and significant variations in preferences between the principals and their agent – yet, had it not been for the 'shirking' of the Commission, the 1992 CAP reform would not have been as far-reaching had the Commission merely acted in accordance with the preferences of Member States.

Eclectic Approaches

The final type of approach to the study of CAP reforms to be considered here is eclectic insofar as it is claimed that different models – some of which are addressed above – are needed to explain the causes of distinct CAP reforms. Along these lines, Ackrill (2000, chapter 5; see also Moyer and Josling 1990) suggests that the establishment of the CAP and the subsequent reforms may be explained through an interest group model, a prominent player's model, an institutional model, or a specified mixture of these. The interest group model, which assumes that governments (in this case including the Commission) respond as a weathercock to interest group pressures, is thought to be able to explain the initial shaping of the CAP in the late 1950s. The prominent players model, which opens up the state and recognises the potentiality of conflicts among state institutions and coalitions across public and private interests, and to an even greater extend the institutional model (see Kay 2000), which downplays the role of interest organisations and instead emphasise the decision-making

---

[3] The Article 113 Committee is a forum made up by Member State representatives where international trade issues are discussed in a dialogue with the Commission. It has now become the Article 133 Committee

competences of various public institutions, are in turn considered to contain greater explanatory value regarding the causes of subsequent CAP reforms.

It is suggested that, whereas a significant degree of the persistency of the CAP seems to be attributed to the activity of the farm lobby both within Member States and among the EU institutions, CAP reforms in 1984, 1988, 1992 and, tentatively, also the CAP reform in 1999 were conditioned by the more or less absence of the farm lobby. Moreover, CAP reforms seem to be conditioned by either budget crisis (the 1984, 1988, and 1992 reforms) or self-imposed budget restrictions (the 1999 reform) (Ackrill 2000, p.106; see also Moyer and Josling 1990, pp.209–211). Like the farm lobby, the Council of Agriculture Ministers tends to stand surety for the continuity of the CAP due to a general reluctance to upset the sensitive pattern of financial transfer and funding of the CAP budget as established during its formative years. However, it is claimed that reform proposals coming out of the Commission are more likely to succeed when – for whatever reason – the level of conflict is high within the Council. Whereas the 1984, 1988 and 1992 CAP reforms have been initiated by various constellations within the Commission and Commission Services, these reforms have attained their momentum through the existence of a high level of conflict within the Council. In both 1984 and 1988, the reform proposals were not initially adopted within the Council but passed on to the European Council to be agreed upon. In 1992 the mere threat of the European Council to interfere with the reform negotiations probed the Council to adopt the so far most radical reform in the history of the CAP. Finally, external pressures such as international trade negotiations (1992 reform) and the enlargement of the EU (1999 reform) are claimed to be increasingly important factors in generating CAP reform (Ackrill 2000, chapter 5).

Summary: Explanations of Continuity and Change

It is commonly agreed that the CAP is characterised by an extraordinarily high degree of continuity. It is also agreed that the reasons for this are largely to be found in the institutional characteristics of the CAP. The explanations offered for the continuity within the CAP are:

1. On a macro level, the fragmented nature of the EU polity means that no single agent holds the political power to carry through reforms and, at the same time, there exists a wide range of players who have the power to veto attempts to reform the CAP (Lenschow and Zito 1998; Daugbjerg 1999; Pollack 2001; see also Fennell 1985, 1987).

2. The CAP is a highly sectorised political field, and is largely resistant to outside influence which may seek to effect a change in the CAP. Hence, the continuity of the CAP is safeguarded since this field is largely

inhabited and governed by a 'club-like' Agriculture Council, the Directorate-General (DG) for Agriculture, the Special Committee for Agriculture and the farm lobby. The latter is both pursuing its interests via EU peak organisations and via national neo-corporatist networks (Swinbank 1989; Grant 1993; Nedergaard *et al.* 1993; Lenschow and Zito 1998; Skogstad 1998; Daugbjerg 1999; Pappi and Henning 1999; Ackrill 2000; Sheingate 2000).

3. Since the farm lobby has successfully exposed the CAP budget to great pressure through its privileged access to political institutions both at the national and EU levels and, since the costs of the CAP are dispersed among European taxpayers, it has proved to be very difficult to prevent CAP expenditures from rising and even much less possible to reduce expenditure (Nedergaard *et al.* 1993).

4. It is claimed that there is a permanent 'free-rider' problem within the CAP. This is because Member States tend to maximise national economic interests by claiming the highest possible returns from the CAP budget were the costs are carried collectively (Nedergaard *et al.* 1993; Ackrill 2000).

5. The development of the CAP has been highly path-dependent. Since its establishment, the CAP has been a central concern of EU politics and the objectives of the CAP were even constitutionalised, that is, written into the Treaty of Rome, which has made it very difficult to introduce change (Skogstad 1998; Coleman and Tangermann 1999).

6. It is claimed that it is a widespread idea that agriculture is exceptional as a sector since agricultural production takes place under unpredictable conditions (weather and imperfect markets), and food production is a matter of national interest. Hence, agriculture is thought to be a concern to be dealt with by public policy and there are legitimate reasons for supporting the sector. Not only is this an ingrained general idea within the agricultural sector itself, it is also claimed to be held by the public at large (Skogstad 1998).

7. It has been suggested that the Commission possesses only limited latitude for pursuing its own preferences for reforming the CAP as it is, during international negotiations, closely monitored by representatives of the Member States in the Article 113 Committee (Pollack 2001). Finally, it is broadly agreed that the Agriculture Council has been very reluctant to reform and is a strong defender of the status quo.

The above points to a series of largely institutional constraints that highly favour continuity over change, yet it is widely accepted that the CAP has in fact been subject to changes. The literature concerning CAP reform draws attention to a variety of causes for this:

1. The GATT/WTO negotiations are seen as being critical for the 1992 reform and in this context the role of the Commission (including the Commission Services) as a policy entrepreneur was particularly important (Patterson 1997; Coleman and Tangermann 1999; Daugbjerg 1999; Sheingate 2000; Fouilleux 2004).

2. It has also been suggested that those who make up the farm lobby have been subject to diversification and are increasingly pursuing differing interests. This is the case both at the European level (primarily due to enlargements) and within Member States (in particular in Germany after reunification), and, hence, it is claimed that the unified pressure from these groups is easing off, or at least that the inactivity of the farm lobby around the adoption of reforms is a favourable condition for change. At the same time, new stakeholders in stabilised international trade relations – such as industrial interest groups – have reported their arrival and, hence, the general interest representation has been diversified, which is also thought to be conducive to change (Nedergaard *et al.* 1993; Patterson 1997; Ackrill 2000; Hennis 2001).

3. Due to the structural changes just mentioned, it is claimed that key individuals have come to enjoy a higher degree of autonomy, which has enabled these skilful individuals to push through changes, examples of whom are the Commissioner for Agriculture, Ray MacSharry, and the German Chancellor, Helmut Kohl, during the 1992 reform (Patterson 1997).

4. The actual and the mere threat of 'outsiders' – including the European Council and Ministers for finance and industry – taking control of not only international negotiations, but also those relating to internal EU agricultural issues, is claimed to have prompted the agricultural policy networks to strive for policy solutions deemed to be acceptable, also to those operating outside these networks (Ackrill 2000; Sheingate 2000). Most recently is has been proposed that the traditionally narrow agricultural networks has open up to include a broader number of interests representing the environment, consumption and rural development: in particular at the national, but also the EU-level (Greer 2005).

5. The actual and the perceived 'negative externalities' of the CAP, such as ever-increasing expenditures, surplus production and the adverse effects of agricultural production on the environment, have, to some degree, redefined the expectations – both inside and outside of agriculture – to what the CAP is to deliver (Nedergaard *et al.* 1993; Patterson 1997; Daugbjerg 1999; Ackrill 2000; Sheingate 2000; Hennis 2001; Greer 2005).

6. It has been suggested that CAP reforms in the 1980s and 1990s have been conditioned by budget crisis. Budget crisis may be generated as a

consequence of pressure created through the actual act of CAP expenditures (the 1984 and 1988 reforms), or by self-imposed budget restrictions (the 1999 reform). Budget crisis may also be caused by an increase in the expenditures of broader EU concerns rather than that of agriculture. Examples include the pressure generated by the financial burden imposed by the implementation of the Single European Act (SEA) in general, the Cohesion Policy in particular and by the enlargement of the EU to the South (the 1992 reform) and to the East (the 1999 reform) (Moyer and Josling 1990; Nedergaard *et al.* 1993; Patterson 1997; Lenschow and Zito 1998; Ackrill 2000). With respect to the issue of a crisis being a condition for change, it has also been indicated, for example by Grant (1997, pp.123–29) and Roederer-Rynning (2003, 2003a), that it may be helpful to consider crisis in broader terms so as to include, for instance, food crisis.

The literature concerning CAP reform provides great insight into the question as to why the CAP is characterised by a very high degree of stability and has shown itself to be very resistant to change, it also points to a number of explanations of why change has taken place. This body of literature, however, also faces a number of challenges in terms of its ability to capture and conceptualise ideational change. The following will put forward five challenges to the literature concerning CAP reform in an endeavour to develop sensitive methodologies and coherent concepts aimed at the study of ideational change. This objective is pursued through a critical discussion of a series of approaches to the study of continuity and change within the CAP.

Five Challenges to the Study of Ideational Change

The first and most basic challenge, which is particularly relevant for network and public choice approaches, is the need to produce, explicitly, a concept that can capture ideational change. For instance, although Daugbjerg (1998, 1999) in his application of a network analytical approach draws on a scale to measure the scope of change cf. Hall (1993) – and it is in fact agreed that changes have taken place – this approach does not explicate a conceptualisation of the dynamics of change. Indeed, as pointed out by Daugbjerg, the existence of cohesive policy networks within the CAP as well as the fragmented nature of the EU polity is expected to stand surety that changes are avoided. Hence, Daugbjerg is, essentially, 'examining the reasons why there have been no fundamental changes in the CAP' (Daugbjerg 1999, p.410). The intriguing questions that subsequently crop up, however, seem to be related to 'how' and 'why' ideational changes have taken place in spite of the existence of cohesive policy networks, and despite the fragmented nature of the EU polity, and regardless of whether such changes are considered fundamental or limited in scope. Likewise,

Nedergaard *et al.* (1993) have a pronounced preoccupation with the persistency of the CAP, but the tentative ideational changes identified in the early 1990s are more difficult to capture through the public choice framework applied. It is suggested that shifts in 'supply and demand behaviour in the political market' have turned agricultural policy into a societal problem. Clearly, however, even in the early 1990s, surplus agricultural production, increasing agricultural expenditures, adverse environmental effects of agricultural production and international trade negotiations were not new phenomena. Hence, if agricultural policy has in fact turned into a societal concern, this development also gives rise to the questions of how and why such a shift took place in the early 1990s as claimed.

A second challenge to the conceptualisation of ideational change is related to empirical sensitivity of the concepts proposed by the literature on CAP reform. A curious observation is that those studies which explicitly apply theoretical frameworks including concepts of change, are also the studies that end up concluding the highest degree of stability. This could mean that either change has not occurred at all or that the concepts and/or methods employed to capture change lack sensitivity. If, however, we accept that – based on the indications indeed made by the existing literature on CAP reform – ideational changes have occurred within the CAP, then the inevitable conclusion seems to be that it is the concepts and/or methods employed to capture change that lack sensitivity. For instance, the term 'shirking' as used by Pollack (2001) in his application of a principal-agent theoretical framework may be seen as a conceptualisation of ideational change.

'Shirking' appears when agents seek to optimise their own preferences at the expense of the preferences of their principals and in that sense agent 'shirking' may contribute to ideational change to the extent that agent preferences are in favour of such. Against this background, Pollack, on the one hand, found that the Commission – as an agent – had only a limited amount of latitude in bringing about change against the wishes of the Member States – i.e. the principals – in the context of the 1992 CAP reform. The reason for this is that the institutional conditions, which the principal-agent theoretical framework suggest as being relevant, all work against the ability of the Commission to 'shirk'. On the other hand, Pollack also concludes – and in accordance with the bulk of the literature on CAP reform – that had it not been for the workings of the Commission, the 1992 CAP reform would not have been carried through so quickly and would not have been as far-reaching. In other words, the principal-agent framework appears to be able to account for the lack of ability of the Commission to bring about change. However, the findings implying that the Commission expedited and enlarged the 1992 CAP reform is not accounted for theoretically. In fact, it seems that the dynamics of ideational change should be attributed to something other than agent 'shirking' as the term is used by Pollack since the institutional conditions deemed relevant all worked to severely constrain potential 'shirking' by the Commission.

Likewise, studies with a focus on policy paradigms or policy frames are not able to identify ideational change within the CAP and common to these approaches are that they have a predefined notion of the ideational developments studied. In fairness, neither Skogstad (1998) nor Lenschow and Zito (1998) explicitly reject the view that the CAP has been subject to ideational change. However, both Skogstad (1998) and – albeit less pronounced – Lenschow and Zito (1998) measure ideational change against analytically pre-defined ideals and pursue the ideals as either/or phenomena. This leaves little room for complexity and possible opposing tendencies and, essentially, fails to reveal the actual ideational basis of the CAP. In fact, these approaches reveal what the CAP is not: the CAP has not been subject to deep-seated liberalisation (Skogstad 1998), and the CAP has not been subject to a significant environmentalisation (Lenschow and Zito 1998). The objection here is not that changes may not have taken place along the lines of the ideational parameters pursued – for instance regarding the state-assistance/ market liberal paradigms as investigated by Skogstad. Rather, the point is that the pursuit of capturing and, essentially, conceptualising ideational change within the CAP is severely confined by analytical strategies, which operate with predefined ideational parameters. Analytical strategies aiming to investigate continuity and change within the CAP are thus faced with a second challenge which involves the development of more inductive methods that are sensitive to the actual nature of the ideational changes that appear to have occurred within the CAP.

A third challenge is related to the theorisation of the workings of policy entrepreneurs. Ideational change is, essentially, a collective phenomenon. However, this does not preclude us from pursuing the identification and conceptualisation of the type of agency, which may give momentum to ideational change. Although the literature on CAP reform frequently points to the Commission and the Commission services as taking the role of a policy entrepreneur and, hence, as being a pivotal agent in bringing about change within the CAP, the concept of policy entrepreneurs is most often not theorised or placed in a broader theoretical framework. When theorised, the momentum given by policy entrepreneurs or 'key' individuals to change tend to depend on the particular skills or psychological predispositions by named individuals or collective agents rather than the conditions which enable the exercise of policy entrepreneurship.

For instance, Patterson (1997) suggests reverberation and synergistic linkages between the international, the Community and the domestic levels of negotiation as conceptualisations of the dynamics of change. However, it is essentially the workings of key individuals (Ray MacSharry and Helmut Kohl), which are critical in carrying through change in regard to the 1992 CAP reform by facilitating links between levels of negotiation. It is suggested that the improved autonomy of key individuals is caused by particular structural changes – a significant increase, in the late 1980s/early 1990s, in the real and perceived costs of not reforming the CAP, and a diversification of interest group pressures (in particular in Germany) – and the structural changes pointed to constitute plausible causes for such increased autonomy. Yet it also appears that, in the end, this type of explanation depends on the particular skills or psychological predispositions of key individuals. Policy entrepreneurship may very well contain an important dynamic of ideational change within the CAP, but the conceptualisations of policy entrepreneurs put forward tend to degenerate into ad hoc, residual explanations of change. Thus, there exists a need to qualify the position or role of policy entrepreneurs so that this concept is not confined to named individuals or collective agents, and avoid theorisation. The need to theorise the conditions for, and quality of, policy entrepreneurship is underlined by recent research indicating that the EP and groupings within the EP – not only the Commission and Commission Services – have given momentum to certain ideational changes within the CAP (Roederer-Rynning 2003).

Fouilleux (2004) continues the attention given to the Commission as a policy entrepreneur, however, Fouilleux also begins to theorise the role of policy entrepreneurship in pushing through ideational change. It is suggested that the exercise of policy entrepreneurship depends on the ability of an agent to use discourse as a resource to give shape to what may be conceived, for instance, relevant and legitimate problems in bargaining processes with other agents. In order for a discourse to catch on and, hence, successfully give momentum to change it must 'fit' into the exiting institutional settings and dominant cognitive frames. That ideas must 'sound right', be 'persuasive', or 'fit' into a particular context in order to be adopted here, is indeed a recurrent notion within the literature concerned with ideational change more generally (e.g. Hall 1993; Hajer 1995). However, while Fouilleux's conceptualisation of policy entrepreneurship is promising, further theorisation is needed to account for the quality of 'fitting' ideas.

A fourth challenge to the study of ideational change within the CAP is related to the externalisation of explanations of change. Sheingate (2000) has a similar preoccupation with perceptual or ideational change as does Skogstad (1998) and Lenschow and Zito (1998), but in contrast to Skogstad and Lenschow, and Zito, Sheingate is able to identify change within the CAP. The understanding of the dynamics of change proposed by the historical institutional approach applied by Sheingate holds, on the one hand, that issue redefinition may lead to venue change and, on the other hand, venue change may bring about new aspects of a particular issue. This dynamic necessitates an external momentum in the sense that 'something' or 'someone' must either trigger the redefinition of issues or, alternatively, trigger a change in venue. Yet, ascribing change to external events is problematic particularly when dealing with the study of ideas. How do we determine whether certain ideas are external or internal to the CAP? In addition, given some ideational changes are more likely than others, what are the conditions for, and through which sorts of processes do ideas travel from 'outside' the CAP to bring about ideational change 'inside' the CAP? A fourth challenge to the literature on CAP reform is to address such questions and, essentially, reconsider the boundaries for the CAP as a policy field. Arguably, the study of ideas also implies a reconsideration of which issues and agents are to be considered to be inside and outside of the CAP as a policy field. For instance, to the extent issues related to the liberalisation of international agricultural trade and enlargements of the EU are part of the ideational foundation of the CAP, such issues – and related agents – may very well be considered to form part of what constitutes the CAP as a policy field. As such, the perceived need to adjust the CAP to the liberalisation of international trade and EU enlargements are endogenous dynamics of change which may effect policy outcomes and do so prior to the actual act of international trade agreements and EU enlargements.

The fifth and final challenge to be put forward here is concerned with the need to pursue the study of ideational change within the CAP over longer periods of time and in-between formal CAP reforms. Reforms, as formally proposed by the Commission and finally adopted by a decision in the Council of Agriculture Ministers, are not necessarily accepted to comprise actual changes by the literature on CAP reform. In fact, most often it is argued that even proposals that start out as attempts to generate genuine alterations of the CAP usually end in (i.e. when the final decision is made by the Council of Agriculture Ministers) significantly watered down versions of the initial proposals. That is, the adoption of a proposal to which the label 'reform' has been attached, is far from being regarded as equivalent to change. Still, there is a strong tendency to theorise over points in time, which have been identified by, most notably, the Commission but also within the Council as constituting a reform of the CAP.

Although a view to formal reform attempts is definitely appropriate in the study of ideational change, an exclusive focus on synchronic cuts in time prevents us from identifying change that may take place over longer periods of time. Moreover, one-sided evaluating reform attempts by comparing the immediate context 'before' and 'after' a point in time, which has been labelled as CAP reform, seems to neglect potential change that may happen 'in-between' reforms. Kay (2003) begins to address this problem employing a path dependency perspective, which emphasises the cumulative effects of, for instance, the introduction of new policy instruments during one reform period on future CAP reforms and policy change. Kay essentially proposes the introduction of a supplementary understanding of change: an understanding, which conceives change as a process (see also Daugbjerg 2003). This notion is particularly relevant for the study of ideational change within the CAP since ideas are unlikely to change with a snap of the fingers, but rather may change over time. In fact, it seems that CAP reforms are the formalisation of ideational change that has been under way for sometimes 10–20 years. As it will be shown, in the context of the 1992 CAP reform, the adoption of environmental regulations for European farming was a concern that had been in the pipeline at least since the mid-1980s – perhaps since the mid-1970s – and subsequently been spreading across various parts of the Commission Services and among EP committees. In other words, a fifth challenge to the literature concerning the reform of the CAP is the development of analytical strategies which enable the study of ideational change over time and in-between reforms, while maintaining a focus on the formalisation of ideas through the adoption of, for example, legal sanctions (which may or may not be identical to the formal adoption of a CAP reform).

It should be noted that when raising the questions of how the type of institutional change that is ideational change in nature may be captured and conceptualised it may be slightly unfair to take a starting point in a body of literature which – pushed to its extremes – is essentially dealing with failed proposals and watered-down reform attempts. Yet this starting point has been especially chosen in order to show that, empirically and theoretically, we already know a great deal about the stability of the CAP. However, it is also indicated that there is still some way to go in the endeavour to develop sensitive and coherent concepts and methods that can be usefully applied in the study of ideational change. In fairness, it should be emphasised that the point here is not that any single approach or study lacks coherence or validity. Rather, taken together and with the objective of capturing and conceptualising ideational change with a focus on the CAP, it can be seen that the existing body of literature on CAP reform lacks the concepts and sensitive methods required to capture ideational change. Moreover, it should also be emphasised that the research questions of the present study owes a great deal to the insights provided by the CAP reform literature. Studies of CAP reforms offer a wide range of explanations as to why this policy field is characterised by a high degree of continuity in terms of politics, polity and policies (as shown by the above mentioned studies). At the same time, ideational changes have increasingly been observed within the CAP (again as shown by the above-mentioned studies) even if coherent and sensitive conceptualisations are rarely offered. Finally, against the background of the CAP reform literature, the CAP offers an empirical opportunity in the endeavour to refine and develop concepts and methods with the aim to capture and, ultimately, explain the type of institutional changes, which are ideational in nature. The CAP is hence seen as a policy field 'least likely' to show institutional changes or, in other words, a critical case (Flyvbjerg 2001, chapter 6) for the study of ideational change.

Outline of the Book

This chapter has argued that the literature concerned with CAP reform is faced by a number of methodological and conceptual challenges in the endeavour to capture and conceptualise ideational and institutional change. Still the nature of institutions and ideas, the interrelation between the two and related dynamics of change needs further consideration and specification. Against this background, the objective of Chapter 2 is to enlarge on methodologies to capture ideational and institutional change and, likewise, enlarge on, and flesh out, the palette of conceptualisations aimed at the study of ideational and institutional change. Chapter 2 will thus carry forward the discussion on conceptualisations and methods to capture ideational and institutional change within the CAP, as indicated above, into the context of new-institutional insights offered from the rational choice, historical and sociological institutional perspectives.

Chapter 3 draws broadly from the insight from the rational choice, historical and sociological institutional approaches to the study of ideational and institutional change, but also seeks to deal with some of the shortcomings of these approaches in the context of a fourth institutional optic – a discursive institutional approach. Chapter 3 will thus argue that the study of institutional change and, in particular, the study of the types of institutional changes, which are ideational by nature, may be advanced by a discursive institutional approach. Moreover, Chapter 3 will present an analytical strategy for the study of ideational and institutional change and prepare the way for an empirical analysis. It will be argued that the articulation and institutionalisation of organic farming within the CAP may be particularly illustrative in terms of the usefulness of the discursive institutional approach proposed as well as illustrative of broader ideational and institutional developments within the CAP in the period from the late 1970s/early 1980s to 2005.

Chapters 4 to 7 is an empirical analysis of ideational and institutional changes within the CAP and takes a particular look at the articulation and institutionalisation of organic farming. The empirical analysis will cover the period from 1968 to 2005, but has an emphasis on the period from the late 1970s/early 1980s onwards. Chapter 8 will conclude on the ideational and institutional changes captured and conceptualised within the CAP by current study and close by pointing to a number of the virtues and limitations of the discursive institutional approach.

2

Institutional Change:
Rational Choice, Historical
and Sociological Perspectives

The boundaries of the new institutional literature are far from being clear cut and there does not exist a complete consensus on the distinguishing features of the various institutional schools of thought and, hence, on the number of approaches within the broad notion of new institutionalism. Peters (1999), for instance, operates with as many as seven institutional approaches. However, most often three schools are identified, namely, (i) the rational choice; (ii) the historical; and (iii) the sociological (e.g. Hall and Taylor 1996). This threefold categorisation will also be the point of departure here.

   Conceptualisations of what makes up an institution vary greatly across the new-institutional paradigm and, hence, what may be seen, for instance, as a change in policy by one type of framework may by others be seen as constituting an actual institutional change. Likewise, variation exists as to whether institutions are studied mainly as dependent or independent variables or as processes. For instance, sometimes particular institutional designs are studied as the product of agents' preferences or broader socio-economic developments (rational choice but also certain historical institutional approaches), sometimes institutions are studied as constraints on political activity and thus shapers of policy outcomes (historical but also certain rational choice institutional approaches) and, yet again, sometimes institutions are seen as being reproduced and possibly changing through processes of collective adaptation and learning (sociological but also certain historical institutional approaches). The rational choice, historical and sociological approaches also differ on the conditions whose presence is considered necessary to enable institutional change to take place as well as on the factors or dynamics that may bring about institutional change (see Campbell and Pedersen 2001a for an overview). The following will thus first address the issues of how rational choice, historical and sociological approaches define and capture institutions: essentially, it will be discussed *what* changes when institutions change and what are the preferred methodologies used to identify institutional change empirically. Thereafter, we will zoom in on a series of conceptualisation and explanations of institutional change with a particular focus on those changes that are ideational in nature.

Conceptualising and Capturing Institutions

From a rational choice institutional perspective institutions may be conceptualised as 'collections of rules and incentives that establish the conditions for bounded rationality, and therefore establish a 'political space' within which many independent political actors can function' (Peters 1999, p.44). Formal electoral and voting systems, various parliamentary systems, legislative rules setting up prohibition and prescriptions for individuals are among the most typical political institutions observed and studied through the rational choice institutional optic. Central to rational choice institutional research are the basic assumptions about individual behaviour. Individuals are thought to behave rationally, strategically and utility maximising with the intent to realise certain interests or preferences, which are most often seen as being exogenous to institutions.

Basically, when utility maximising individuals go about their daily lives they will, every now and then, be confronted by some sort of institutional set up which constrains their ability to act freely. As individuals obtain knowledge about the working of the institutions confronting them, they will adjust their behaviour accordingly with the aim of maximising their utility within the given constraints. For instance, when members of the Council of Ministers seek to reach a decision, they most often do so under, either the institutional constrains of unanimity, or QMV, which, depending on the issue at hand, is the legally required support needed to make a binding decision. The ministers involved – given the presence of either of the above-mentioned institutional constraints – are expected to act strategically by, for example, entering coalitions and putting forward policy amendments which best serve them in terms of maximising their preferences. In other words, the utility maximising behaviour of political agents is constrained by the institutional arrangements that confront these agents and policy outcomes are shaped by the interaction between agents and institutions.

The reason why institutions arise in the first place is related to the need to deal with problems of collective action. When utility maximising individuals act in concert, problems of collective action may arise such as free-riding, rising transactional costs and uncertainty about the behaviour of other agents (Olsen 1971 [1965]). Political agents may thus find that their goals are best achieved through the creation of certain institutions that constrain them and, importantly, constrain other agents involved, into following certain basic rules of behaviour (Peters 1999, pp.43–45). Downing (1994, p.115) argues that the explanatory value of rational choice institutionalism lies in the fact that the models developed within this approach are 'realistic enough', on the one hand, to capture the specificity of a particular institutional arrangement and, at the same time, still enable general theoretical statements and the application of general models across comparable cases. As such, it is the basic assumptions of individual utility maximisation and the empirical focus on actual 'mass behaviour' combined with the commitment to elaborate on general models that grant the rational choice institutional perspective its explanatory advantages.

This also implies that the preferred methodology of rational choice institutionalists goes through hypothetical-deduction. The point of departure for explaining the origin and function of a given institution will refer back to the basic assumptions about individual rationality. The rational calculation and utility maximisation of individuals are given a priory and on that background the function and benefits generated from a given institutional arrangement may be deduced (Hall and Taylor 1996, p.945). Knowing that individuals act rationally, a particular institutional set up may be explained by the benefits obtained through, for instance, a reduction of transaction costs or the elimination of free-riding and the functioning of such institutional arrangements will depend on, for instance, a shared commitment to some type of commonly accepted enforcement mechanism. Basically, drawn from the assumptions about individual rationality, institutions are established intentionally and individuals enter voluntarily into institutional arrangements with the aim of maximising their utility.

Within the historical institutionalism, institutions are seen as 'formal or informal procedures, routines, norms and conventions embedded in the organizational structure of the polity or political economy' (Hall and Taylor 1996, p.938). The institutional conventions observed from a historical institutional perspective are, for instance, formal voting rules and procedures of consultation with interested parties, informal practices of negotiation and bargaining, bureaucratic routines, prevailing norms and conventions guiding the formulation of policy programmes or the activities of agents within a particular policy field. Historical institutionalism is thus very inclusive as to the type of structural features considered to have institutional characteristics and effects. Essentially, however, institutions are considered to constrain and structure political activity and policy outcomes. On the one hand, institutions shape political interaction, preferences and strategies and, on the other hand, have a direct impact on policy outcomes (Thelen and Steinmo 1992, p.9). The nature of institutions is often related to the organisational particularities of national polities or, in the case of the EU, the structural characteristics of the EU polity.

In order to illuminate such organisational particularities comparative studies have been central to historical institutional research. It may take the form of comparing policy sectors across countries or variations and similarities in national polities. Moreover, broader contextual factors are often prominent in historical institutional studies and, although the historical institutionalism has a rich theoretical agenda from which hypotheses are deduced and tested, research strategies ascribing to this approach will often have a significant inductive element (Campbell and Pedersen 2001a, p.12; Thelen and Steinmo 1992, p.12). Although hypothetical-deductive research designs are not rejected, historical institutionalists tend to pursue an interest in illuminating the complexity of politics through case studies (or comparative case studies) which include studying a vide range of potential explanatory variables. Consequently, explanations proposed by this approach are often multi-dimensional and have a significant degree of historical and contextual specificity.

The broad notion of sociological institutionalism, which is the label most often used to group the institutional approaches that are neither rational choice nor historical institutional, can be seen as containing two sub-branches. One which tend to emphasise the normative dimention of institutions and one that tend to emphasise the cognitive dimension of institutions (Hall and Taylor 1996, p.948). March and Olsen (1989) represent the normative dimension and – as summed up by Peters (1999) – they see institutions as 'a collection of values and rules, largely normative rather than cognitive in the way in which they impact institutional members, as well as the routines that are developed to implement and enforce those values' (Peters 1999, p.29). Values, rules and routines embedded in institutions are thought to shape the behaviour of individuals committed to these institutions by establishing a set of norms for legitimate and appropriate behaviour. Hence, unlike the rational choice institutional line of thinking, the formation of preferences are endogenous to institutions and rather than pursuing the maximisation of individual material utility, agents are guided and motivated by a 'logic of appropriateness'. When agents act according to the logic of appropriateness they will be 'fulfilling the obligations of a role in a situation, and so of trying to determine the imperatives of holding a position. Action stems from conception of necessity, rather than preference' (March and Olsen 1989, p.161).

On the more cognitive side we have the sociological institutionalists, sometimes also referred to as organisational institutionalists (Campbell and Pedersen 2001) or social constructivists (Schneider and Aspinwall 2001). The sociological institutionalist emphasising the cognitive dimention of institutions see institutions as 'cultural rules giving collective meaning and value to particular entities and activities, integrating them into larger schemes' (Meyer *et al.* 1994, p.10). The key words here are cultural rules and collective meaning. Although critique of the blurred lines of demarcation between the notions of culture, socialisation, institutionalisation and the concept of institution has been voiced both from within the sociological institutionalism itself (Jepperson 1991, pp.147–150) and by outside observers (Hall and Taylor 1996, pp.947–948; Peters 1999, p.97), it is clear that this cognitive view on institutions is concerned with the processes through which collective meaning is produced. By suggesting that institutions represent collective meaning systems, institutions have become the optics through which institutional members interpret and 'make sense' of the social world. In that way institutions 'do not simply affect the strategic calculations of individuals...but also their most basic preferences and very identity' (Hall and Taylor 1996, p.948) or, in other words, preferences and identities are endogenous to institutions. Peters (1999, p.103) describes the distinction between a normative and a cognitive view to institutions as subtle and suggests that:

[t]he cognitive view may be more basic than the normative view, given that it determines how the member of the institutions interprets data from the environment, while 'all' the normative perspective tells him or her is what the appropriate behaviour would be in any situation.

Even if subtle and sociological institutional research strategies often seem to cut across the normative/cognitive distinction, it is important to keep this distinction in mind – see below. Regardless whether emphasis is put on the cognitive or normative dimension, however, it follows from the conceptualisations of institutions that the object for study among sociological institutionalists is not individual behaviour but rather the norms that shape preferences, strategies and the assignment of certain roles to individuals and the systems through which agents interpret and give meaning to the world. Although historical institutionalism over time has developed a richer theoretical agenda, sociological institutionalism has from the outset – as is the case with rational choice institutionalism – had a pronounced theoretical ambition. Drawing broadly from sociological theorising, the preferred methodology of sociological institutional studies has been hypothetical-deductive. Yet, a recent interest in moving towards more inductively constructed and historically sensitive analytical strategies seems to have developed (Campbell and Pedersen 2001a, p.12, 2001b, p.253). Altogether, across the rational choice institutional, historical institutional and sociological institutional schools of thought great variation exists as to how institutions are conceptualised and captured. That is to say, a great deal of variation exists as to the question of *what* is changing when institutional change come about as well as on the methodological settings preferred in the pursuit of capturing institutional change. In this light we will now turn to a discussion of a number of unlike takes on the conditions for, and dynamics of, institutional change with a particular focus on those that are ideational in nature.

Conceptualising and Explaining Institutional Change

To be sure, the issue of institutional change has not been a central concern to the rational choice institutionalism and much less so has ideational change. Yet, in general, like decisions to establish and commit to institutions are based on rational and calculated choices, so is the choice to change institutions. If a certain institutional arrangement is considered to have failed to meet the requirements of the individuals involved, then it is expected that a decision to change or perhaps eliminate that institutional arrangement will be made (Peters 1999, p.56). The methodological individual starting point for rational choice institutionalists thus implies that changes are basically related to individual choices. As formulated by Downing: '[I]ndividuals *cannot be taken out* of such explanations [of social life], and their actions are the *cause* of social change. This latter clause does not contradict the view that actions are shaped or *structurally suggested* by the relationships and institutions in which individuals are placed' (Downing 1994, p.110; original emphasis). Likewise, North (1990, p.5) argues that since institutions are human constructs and are changed by humans, then a theory of institutional change must take its point of departure in the individual. From this starting point the rational choice institutional perspective *do* offer conceptualisations and explanations of institutional and ideational change related to shifts in costs and benefits and to the working of policy entrepreneurs.

In spite of the pronounced focus on institutional path-dependency (see below), advocates of historical institutionalism have also developed an interest in ideational and institutional change – and, over time, increasingly so. This interest is reflected in some concern with political learning processes, the power and 'fit' of ideas in politics and crisis as a condition for change. Concepts and explanations of ideational and institutional change are from the sociological institutional viewpoint often related to some sort of learning processes where processes of diffusion and isomorphism may be seen as referring to more specified types of learning. From the literature, which is often associated with policy analysis but still having clear references to, in particular, the normative emphasis of the sociological institutionalism, it has been suggested that processes through which problems are identified, defined and redefined contain a dynamic of change. This latter type of literature also make some suggestions about the nature of 'windows of opportunity', which are points in time where conditions conducive to changes are pronounced. Finally, this optic also has a view to how the working of policy entrepreneurs may be significant in carrying through change. It is these issues that will be addressed further.

Costs and Benefits

Rational choice institutionalists touches upon the potential dynamic of ideational and institutional change only rarely and tends to associate the force of ideas with revolutionary changes or as being, in the main, a relevant variable in marginal cases involving strong moral issues (North 1990, pp.84–86, 89–91). Yet, even if the issue of institutional change and much less the issue of ideational change is only of marginal concern to the rational choice institutionalism, it is possible to tease out considerations on the role of ideational factors in bringing about institutional change from theories with a point of departure within this tradition.

The central dynamic of change proposed by rational choice institutionalism is thus related to shifts in costs and benefits among political agents. Shifts in relative prices are thought to alter the incentive structures constraining human interaction and considered the most common cause of institutional change. Institutional change is thus based on choices made by individuals on the background of a complex weighting of the expected costs and benefits of a proposed institutional change compared to the status quo institutional arrangements (North 1990; Ostrom 1990). Not only is the cost/benefit analysis made complex by the variety of factors to consider and compare, this complexity is further increased by the fact that the weighting of costs and benefits are carried out against a background of human judgement. That is, whereas shifts in relative prices is the dynamic identified as the basic source of institutional change, in particular North (1990) suggests that *perceptions* of shifts in relative prices may vary according to the ideas ascribed to, and the information possessed by, the involved agents. That is, individuals will acquire certain knowledge about shifts in the relative price of, for example, available technology, and process this knowledge through 'pre-existing mental constructs' and make their institutional decisions accordingly (North 1990, p.85). Likewise, when deciding whether to initiate institutional change or not, individuals will consider the expected costs related to the transformation of existing institutions and, in particular, the up-front costs of transformation and the costs of monitoring and enforcing the proposed institutions. The costs related to transformation will, among other things, be low if there exist a shared norm of restraining opportunistic behaviour among the involved agents and the cost of monitoring and enforcing institutional change will, among other things, depend on the degree to which the proposed changes are perceived to be legitimate among the involved (Ostrom 1990, pp.202–205). On the one hand, a shift in relative prices is the prime condition for institutional change to occur and, on the other hand, agents' perceived costs and benefits are central to the institutional decision taken or, in other words, a central dynamic institutional change.

According to North (1990, p.86), institutional changes are most often incremental and it is implied that formal, i.e. contract-based, institutional changes are most often followed by more informal, i.e. norm-based, institutional change (not the other way around). It is further implied that the speed at which informal institutional change comes about is slower than that of formal institutional change (I will return to the rather crude distinction between formal and informal institutional change below). Importantly, not any type of institutional change is equally likely to occur, rather the '[p]ast institutional choices open up some paths and foreclose others to future development' (Ostrom 1990, p.202) (see also North 1990, chapter 11). This latter point – the concept of institutional path-dependency – is shared with the historical institutional approach and implies that power relations and institutional decisions taken around the establishment of a given institutional arrangement are significant for future institutional and policy developments.

Learning

Learning as a dynamic of ideational and institutional change has both been theorised within the historical and sociological institutional line of thinking. Whereas learning processes as a specific description of processes of socialisation tend to be a basic feature of sociological institutional approaches, the emphasis put on learning varies across the historical institutionalism. Historical institutional approaches thus varies from those placing little confidence in learning processes to appear in politics (Pierson 2000), over those characterising politics in terms of continuity and minor adjustments but which also leaves room for shorter periods of change through learning (Baumgartner and Jones 1993), to those stressing learning as a key dynamic of change (Jenkins-Smiths and Sabatier 1993). [4] Those who place little confidence in learning in politics also tend to be those who put the strongest emphasis on the concept of path dependency. This central concept refers to the significance of past institutional decisions for the structuring of political activity and future policy outcomes. Periods of institutional creation are highly important since the political context around these points in time is thought to have lasting effects on future developments. Prevailing power relations, preferred administrative and political processes as well as norms and ideas among agents during times of institutional creation will become embedded in these institutions and set out a path for the future.

Pierson (2000) suggests that the reason why path-dependency is a central feature of politics can be found in the dynamic of increasing returns. The belief is that '[i]n an increasing returns process, the probability of further steps along

---

[4] Although Jenkins-Smiths and Sabatier (1993) seek to dissociate themselves from institutional theory, their concept of policy-oriented learning have found its way to historical institutional research.

the same path increases with each move down that path. This is because the *relative* benefits of the current activity compared with other possible options increases over time' (Pierson 2000, p.252; original emphasis). Asymmetrical power relations among political agents thus not only become embedded in institutional arrangements around their creation, political authority and asymmetrical power relations are constantly being reproduced and tend to be reinforced as time passes (Pierson 2000, p.259). From this perspective, it is unlikely that changes may come about through learning processes since such processes are rare in politics due to the complex nature of political goals and the weak link between political action and outcomes (Pierson 2000, p.260). This is particularly so in an EU context since politics here is characterised by a very high degree of complexity due to, for instance, the vast number of political agents involved and the high density of potential issues to be dealt with (Pierson 1998, pp.39–40). These basic features of politics and institution building work in favour of a very high degree of political stability and the notion of increasing returns suggest that the firmness of a political equilibrium will increase over time. Essentially, when certain political goals and courses of actions have been institutionalised in formal rules and procedures, internalised in political culture, identity and belief systems, departure at this point is almost impossible.

Even if the basic features of path-dependency is upheld, other historical institutionalists makes room for ideational and institutional change through learning or feed-back processes by the notion of punctuated equilibrium. Along the lines of path-dependency, institutions will for long periods of time exhibit a high degree of stability and ensure stable power relations and policy outcomes. During these periods of time, limited, incremental and reversible changes may occur. Incremental change may come about as decision-makers may want to adjust for the unforeseen consequences of decisions taken during the creation of a given institutional arrangement or change may be caused by a counter mobilisation by groupings, which find themselves adversely affected by the current course of action. Either way, the likeliness of incremental change decreases as time passes since the costs involved in political struggles tends to increase, since the more powerful groupings will only give up so much power, and since the less powerful groupings will soon reach the limits for improvements possible within a political system characterised by equilibrium (Baumgartner and Jones 1993, p.9, 16). The punctuated equilibrium line of thinking, however, also includes suggestions as to how a political equilibrium may under certain conditions be punctuated and, after a period of rapid change, a new political equilibrium may appear. That is, the concept of punctuation seeks to conceptualise rapid and radical institutional change. Baumgartner and Jones (1993) suggest the dynamic of institutional change has to do with processes of positive feed-back. Positive feed-back processes may involve the diffusion of ideas between policy sectors or across political systems and, as these ideas gain increased support both among the previously marginalised as well as the more powerful groupings, the political equilibrium may be punctuated making room for radical change. Issue definition and redefinition is of significant importance here. Whereas clear and stable definitions of the issues with which a political system is concerned contribute to the stability of this system, a challenge and possible redefinition of the prevailing political issues may lead to radical changes. The particularity of a given political institutional structure, such as accessibility to a political system and possible changes in these institutional structures, will be decisive for the likelihood of an issue to be defined in a new way and affect the potential mobilisation of support backing a newly defined issue (Baumgartner and Jones 1993, pp.15–21). In turn, issue redefinition may bring about venue change. If an issue finds a new expression it may cause that issue to be elevated to another forum where other political agents may deal with it in new ways (Baumgartner and Jones 1993, p.37). The basic dynamic of change suggested by the concept of punctuation is thus one emphasising the interaction between, on the one hand, the definition and redefinition of issues and, on the other hand, the venues where these issues are dealt with. Change in this interaction seems to depend on the introduction of change from external sources.

Those giving the most attention to learning processes among historical institutionalists are also those giving the highest degree of attention to the ideational. From this perspective, policy-oriented learning has been suggested as a dynamic that may give momentum to changes in beliefs systems which in turn form the basis for changes in policy outputs (Jenkins-Smiths and Sabatier 1993). Policy-oriented learning may come about against the background of, for instance, individual learning and change in attitude, the diffusion of new beliefs and attitudes among individuals, or a group dynamic involving conflicts among groups or a polarisation of groups (Jenkins-Smiths and Sabatier 1993, p.42). Technical information concerned with the performance of policies may also potentially illuminate gaps between policy goals and policy outcome or even question prevailing causal assumptions of policy programmes that, in turn, may lead to the readjustment of belief systems. Finally, by engaging in an analytical debate, supporters of a deprived belief system may challenge the validity of a proposed and favoured policy objective, the causal assumptions behind a policy programme and, also, the efficacy of the institutional set up surrounding a given policy (Jenkins-Smiths and Sabatier 1993, p.45). That is, opposed to a one-sided focus on the lasting effects of the political situation around the creation of new institutions, change is also possible at later points in time if novel technical information sees the light of the day, if the validity of the prevailing belief system is challenged and, essentially, if the structure of existing belief systems is changed through learning processes. It should be noted that this type of learning processes may lead to change in the policy positions taken up by the involved agents and the strategies employed to attain these preferences, but policy-oriented learning is very unlikely to alter more fundamental normative and ontological axioms (Sabatier 1993, p.35). In the unlikely case that change does appear in the more fundamental normative and ontological axioms of a belief system, it will be caused by external system-wide events such as change in socio-economic conditions or government. The types of changes that may be given birth to by policy-oriented learning processes are thus ideational and rarely radical.

Whether processes of learning are taking place among individuals, within organisations, or across organisational fields,[5] learning is considered to contain a central dynamic of institutional change within sociological institutional thinking. Along these lines, learning involves 'an accomplishment in terms of improved knowledge, skills, performance, and preparedness for the future' and learning processes are thought to have occurred 'when observations and inferences from experience create fairly enduring changes in organizational structures and standard operating procedures' (Olsen and Peters 1996, p.6). Learning, however, is not so much a process involving rational calculations on a series of institutional alternatives and, in consequence, the adoption and implementation of those institutional changes that anticipate future developments in the best possible way. Rather, and along the lines of 'the logic of appropriateness', lessons learned from experiences are turned into changes in organisational structures and routines through the matching of procedures to situations.

Organisations are thought to learn by encoding inferences from experiences into organisational routines. Hence, lessons from experiences are institutionalised and passed on among organisational members as well as becoming available to other organisations through 'socialization, education, imitation, professionalization, personnel movement, mergers, and acquisitions' (Levitt and March 1988, p.320). Two central conceptualisations of processes of learning within the sociological institutional tradition are diffusion and isomorphism.

Diffusion, in general, refers to processes where the spread of norms to a wider audience leads to the internalisation of new sets of understandings about what is considered appropriate in certain contexts. Such processes come about as still larger groups are socialised and learn the appropriateness of the norms being diffused and come to act accordingly (Checkel 2001, p.30ff.). In other words, the successful diffusion of norms is a process of institutional change. Whereas diffusion is a conceptualisation of processes between social contexts or organisational fields (Strang and Meyer 1994), institutional change driven by the

---

[5] Scott (1994, p. 71) defines organisational fields as 'communities of organizations that participate in the same meaning systems, are defined by similar symbolic processes, and are subject to common regulatory processes'. DiMaggio and Powell (1991) supplement this definition by considering an organisational field as being made up by 'those organizations that, in the aggregate, constitute a recognized area of institutional life: key suppliers, resource and product consumers, regulatory agencies, and other organizations that produce similar services or products' (DiMaggio and Powell 1991, pp.64–65).

copying of norms and practices among organisations within organisational fields is termed isomorphism and tends to proceed regardless of whether or not organisational efficiency is enhanced (DiMaggio and Powell 1991). Sociological institutionalists tend to suggest that norms and social practices are diffused and copied between and across organisational fields in a search of success and legitimacy at times of uncertainty about organisational goals, technological causes and effects, and external expectations and that, as these processes proceed, institutional practices and norms become more homogenous (DiMaggio and Powell 1991; Strang and Meyer 1994; Olsen and Peters 1996). On the one hand then, organisational structures and routines embedded in specific organisations may be subject to change in the search of success and legitimacy. On the other hand, it may also be argued that, overall, processes of diffusion between social contexts or isomorphism within organisational fields are processes of continuity and homogeneity rather than change and diversification. Whereas the historical institutional conceptualisation of learning processes tends to focus on the exchange of viewpoints through bargaining, conflicts and 'analytical debates' between various groupings, processes of learning, diffusion and isomorphism seem – from the sociological institutional perspective – to be more discrete processes of socialisation.

The Power and 'Fit' of Ideas

As it has already been hinted at, the study of the role of ideas has attracted a great deal of attention within the historical and sociological institutional tradition and less so within rational choice institutionalism. In particular among historical institutional approaches certain ideas are, on the one hand, often attributed with a persuasive power in themselves, which may bring about institutional change. On the other hand, both among historical and sociological institutionalists, the effect of ideas on institutional change tend to depend on the 'fit' between the ideas awaiting to be introduced into a particular social context and the ideas already embedded in this context.

Both from within the historical institutional and sociological institutional line of thinking is has been suggested that electorate power struggles, shifts in public administration or government, and interest group pressure are sources of change. Yet, both stances give significant attention of the role of ideas in bringing about change. From a historical institutional perspective Hall (1989), for instance, has observed that ideas:

had the power to change the perception a group had of its own interest, and they made possible new courses of action that changed the material world itself. In these respects...ideas had a good deal of independent force over circumstances' (Hall 1989, p.369). Moreover, '[w]hen an evocative set of ideas are introduced into the political arena, they do not simply rest on top of the other factors already there. Rather, they can alter the composition of other elements in the political sphere, like a catalyst or binding agent that allows existing ingredients to combine in new ways' (Hall 1989, p.367).

The dynamic of change suggested by Hall is related to the ideational interlinks between policymaking and society at large and it is described as a process of social learning through which policy paradigms may be subject to change. A change in the policy paradigm informing a particular policy field is thought of as a shift in the interpretative framework through which policymakers formulate policy goals, consider the type of instruments that should be employed to realise these goals and identify the problems that ought to be subject to public policy (Hall 1989, 1993). Hence, a policy paradigm is seen as a coherent set of ideas that delimit the types of policy problems, instruments, and goals that authoritative policymakers consider legitimate from those, which are judged as illegitimate. Essentially, the dynamics of change brought about by ideational factors has to do with the persuasive powers of a given set of ideas and how the ideas being diffused fit with already existing ideas and institutions.

The latter conception is shared by sociological institutionalists, yet, more than historical institutionalists, approaches within this tradition imply that ideational and institutional changes are largely unpredictable and random (Cohen *et al.* 1972; Kingdon 1995).[6] This perspective takes its point of departure in a 'Garbage Can Model' of decision-making as first introduced by Cohen *et al.* (1972). This model was designed to deal with decision-making in 'organised anarchies' characterised by inconsistent and ill-defined preferences, unclear decision-making structures and fluid participation. Although, at the time Cohen *et al.* (1972) saw organised anarchies as a marginal phenomenon, it subsequently became increasingly common to identify decision-making as complex processes, which are rarely guided by the narrow rational material utility maximisation by individuals. The central point is that decisions are not primarily made with the intent to solve problems against a background of rational calculations. Rather, choices are accidental and appear at times when just the right combination of problems, solutions and decision-makers coincide within an organised anarchy. This line of thinking implies that important institutional choices often appear by accident and is highly dependent on the particularities of a given organisational structure or polity.

Accordingly, Kingdon (1995) points to three distinct processes concerned with problem identification, policy formulation and politics. The three processes of problem identification, policy formulation and politics are thought to evolve independently of each other and, only rarely, and largely unpredictably, run together. However, when they do run together the possibility for policy

---

[6] Although the institutional perspective of Cohen et al. (1972) and more so of Kingdon (1995) have receded into the background, the normative emphasis may well allow their inclusion in the broad sociological institutional approach.

initiatives to be taken and changes to appear, increases dramatically. A window of opportunity will most often open due to the identification or redefinition of a problem or due to changes within the political process. Attention is drawn to a new or redefined problem or a shift may appear in governmental or administrative personnel or in the national mood. Such changes provide the opportunity for an issue to be forwarded to a decision agenda with a highly increased possibility of being considered favourably. Yet whereas such opportunities may in some instances be predictable most often they are not. That is, an important dynamic of change from this perspective is related to the processes through which problems are identified, shaped and selected for political consideration. Political activity is seen as having as much to do with 'pet-solutions' being attached to those problems considered legitimate at any given time, as with intentional problem-solving activities.

Policy Entrepreneurs

The final type of explanation of institutional change to be considered is related to the concept of policy entrepreneurs. This type of explanation has both a rational choice and sociological institutional representation whereas historical institutionalists have been more reluctant to theorise the contribution of agency in bringing about change. From a rational choice institutional perspective institutions are seen as human constructs that set up certain constraints on the behaviour of political agents which, nonetheless, are utility maximising creatures. However, the concept of political entrepreneurs introduces into rational choice institutionalism the possibility of certain individuals of not acting with a narrow utility maximisation concern in mind. That is, individuals may be ideologically committed to particular points of views and thus willing to make sacrifices, which defy narrow cost–benefit analysis (North 1990). Accordingly, Downing (1994, p.106, 115) attaches a particular importance to 'key actors', who may not always be guided by a narrow logic of egoistic utility maximisation. Rather, the actual beliefs, desires and action of key actors should be subject to empirical investigation as such knowledge may contain essential explanatory power for particular institutional decisions. This line of thinking is neither new nor unusual and there exist at least two types of explanations of the working policy entrepreneurs in rational choice institutional theorising. One type of explanation tends to emphasise the 'inner directions' of certain agents and another tend to explain the ability of certain agents to bring about change with reference to their position in a broader institutional environment.

Combining the two types of explanations, Downs (1967) singled out particular types of individuals as significant for the development of policies. Downs identified three types of officials, namely, zealots, statesmen and advocates, all of whom were seen to stand out as officials driven by 'mixed motives'. In addition to the maximisation of particular self-interests, such bureaucrats will have a certain concern for specific policy goals (zealots), very broad policy goals (statesmen) or the goals of the organisation or unit they inhabit (advocates). The behaviour of these individuals is explained partly by psychological predispositions (optimism, energy, aggression) and partly by the behavioural requirements inherent in the positions held by these individuals. Latter rational choice institutional theorising on the working of policy entrepreneurs tends to give primacy to the types of explanations, which emphasise the positions held by certain agents in the broader institutional environment. On the one hand, the role of policy entrepreneurs in bringing about institutional change is related to their contribution to evaluating the costs and benefits of alternative institutional arrangements (North 1990; Ostrom 1990; Doron and Sened 2001). The evaluation made by policy entrepreneurs depends on their ability to generate knowledge and reduce uncertainty about alternative institutional arrangements and policy outcomes, and suggest the more profitable alternative. Knowledgeable agents may thus contribute to the shaping and altering of existing perceptions of institutional costs and benefits relative to alternative institutional arrangements and thus bring about institutional change (North 1990, p.29, 87; Doron and Sened 2001). On the other hand, the ability of certain agents to generate knowledge and in turn alter the perceived costs and benefits of alternative institutional arrangements is related to the position held in the institutional environment. It may be emphasised that by operating at the margin of the existing incentive structures, policy entrepreneurs may require knowledge about alternative courses of action and may identify opportunities for profitable institutional change (North 1990, p.100).

Alternatively, policy entrepreneurs may be seen as operating as mediators between constituencies and government officials and from this position generate and mediate information about the intended action of constituencies, and the cost and benefits of the actions of both government and constituencies. Policy entrepreneurs may thus be influential on government action insofar that their involvement is considered an important indicator of the scope of support behind the issue advocated by the political entrepreneur. If a political entrepreneur decides to carry the costs of advocating a particular interest, potential beneficiaries and government officials are believed to see that as a sign of the existence of a viable group of support. Similarly, if political entrepreneurs decide not to carry the costs of advocacy, then potential beneficiaries and government officials will interpret that as a lack of support within the constituency (Doron and Sened 2001, pp.77–78, 87). Basically, the introduction of policy entrepreneurs as potential generators of change is also the introduction of a particular kind of agency. The kind of agency that may be significant in carrying through institutional change through a given combination of extraordinary skills and the position upheld by the policy entrepreneur in the broader institutional environment.

Theorising on the working of policy entrepreneurs are also found with the sociological institutional tradition – particularly the normative variant hereof. As is the case with the rational choice institutional approach, the concept of policy entrepreneurs is a sociological institutional attempt to specify a particular kind of agency: the type of agency that may contribute to carrying through change. Compared with rational choice institutionalism, sociological institutional theorising has the strongest emphasis on institutional structures vis-à-vis the role of agency. There are, however, important variations within the sociological institutional optic related to whether emphasis is put on the normative or cognitive institutional dimension. The cognitive emphasis seems to have had very little concern for agency due to the view that institutions both shape the preferences and identities of agents as well as provide the framework through which these agents interpret the world. The normative emphasis, on the other hand, upholds the view, regardless of the degree of institutionalisation that a particular set of values may have attained, that these values are never unambiguous and all embracing and institutional members are thus assigned with a certain space for interpretation (March and Olsen 1996, p.251ff.; Peters 1999, p.29). Furthermore, the complexity of the institutional environments that agents often find themselves situated within implies that there may be some leeway for agents to determine the appropriate behaviour in a given situation: hence, 'the fact that most behaviour is driven by routines does not, by itself, make most behaviour routine' (March and Olsen 1989, p.24). For these reasons, it is also within the body of the sociological institutional literature that has a normative emphasis, where reflections on the working of policy entrepreneurs are to be found.

From this perspective, policy entrepreneurs has been seen as possessing a 'willingness to invest…resources – time, energy, reputation, and sometimes money – in the hope of a future return' and they may be found anywhere among the people involved in policymaking (Kingdon 1995, p.122). That is, policy entrepreneurs are supplied with a certain 'willingness', which rises out of certain expectations about 'future returns', and they are attributed with particular capabilities that enable engagement in generating change. The capabilities of policy entrepreneurs may have to do with their 'expertise and knowledge in their given field; substantial negotiating skills; persistence; connections to relevant political actors' (Checkel 1997, p.9).' Policy entrepreneurs will be particularly active around points in time characterised by a search for alternatives or when a crisis is perceived to exist. These points in time (cf. windows of opportunity above) will provide the opportunity for policy entrepreneurs to forward particular ideas or norms (Kingdon 1995, p.166, 203) and attach their ready-made solutions to a problem to which attention has been drawn. On the one hand, policy entrepreneurs are attributed with certain individual resources and qualities. On the other hand, the ability of policy entrepreneurs to use their resources and qualities in favour of the spread of particular ideas and norms is conditioned by an 'open' window of opportunity, just like their success is significantly enhanced if the ideas and solutions advocated fit the problems recognised by powerful political leaders and organisations.

Crisis as a Condition for Change

The existence of some sort of crisis is often pointed to as conducive for ideational and institutional change across the rational choice, historical and sociological institutionalisms, but crisis as a condition for change has been given particular attention by historical institutionalists. Regardless whether the dynamics of change are conceptualised along the lines of, for instance, punctuated equilibriums, various types of learning processes, or the independent role of certain powerful ideas, a situation characterised by crisis is generally considered conducive to the emergence of institutional change (e.g. Hall 1989; Baumgartner and Jones 1993; Sabatier and Jenkins-Smith 1993; Olsen and Peters 1996). Within the historical institutional tradition, crisis is often seen as an objective, non-cognitive and external condition for change. That is, a crisis tends to arise against the background of what is thought of as non-cognitive factors and most important in this regard are general economic recessions or even acts of war. Moreover and related, crisis is often seen as being triggered by factors external to the policy or polity studied. Whereas a situation characterised by crisis is conducive to incremental institutional change, it seems that the existence of a crisis is close to being thought of as a precondition for radical institutional changes.

Regardless of the sources identified as triggering a situation of crisis, historical institutional attempts to explain institutional change will often involve a reference to the necessity to address an extraordinary political situation against which the current institutions are incapacitated. A crisis will thus often be related to the failure of policy programmes or may be related to the accumulation of political and economic contradictions. However, some theorists operating at the margin of the historical institutional tradition also make room for crisis as an ideational phenomenon where a crisis involves a perceived need to make an intervention (Hay 2001). Still, it should be noted that crisis as a frequently referred to condition for change is thought of and studied quite differently from the dynamic of change found in those sociological institutional approaches concerned with the processes of problem identification and shaping and for which changes are more random. Among sociological institutionalists, a crisis may be thought of as an objective and non-cognitive phenomenon, but most often a crisis is seen as an ideational condition conducive for institutional change (e.g. Olsen and Peters 1996).

Summary: Rational choice, Historical and Sociological Perspectives on Institutional Change

The search for concepts and explanations of institutional change and, particularly, the type of institutional change which is ideational in nature, has had a bias towards the dynamics of, and conditions for, change indicated in the CAP reform literature, but also enlarged on the methodologies and concepts offered. In order to promote analytical clarity the search for concepts and explanations of institutional change was conducted along the lines of the commonly used three-fold categorisation: rational choice institutionalism, historical institutionalism and sociological institutionalism. Thankfully, more often than not the proponents of a certain institutional optic have overtly drawn inspiration from other parts of the new-institutionalism and, hopefully, this point has also been be discernible above. However – with this in mind – in the name of further analytical clarity Table 2.1 emphasises the differences between the three institutional schools of thought, rather than focus on the details, overlaps and nuances. In other words, Table 2.1 establishes three ideal-typical approaches to the study of institutional change. The search for conceptualisations and explanations of institutional change is summed up along the lines of the methods employed to capture institutional change, the conditions thought of as conducive to institutional change as well as the central dynamics [7] through which institutional change is brought about.

As to the dynamic of institutional change, the type of change that the dynamics are associated with is indicated in brackets and described by the terms formal, informal, radical or incremental change. These categories are, it must be acknowledged, quite crude and their usage across the theoretical frameworks are not straightforwardly comparable. Nonetheless, they represent the most commonly used terms that describe unlike types of change. Formal institutional change seems to imply, for instance, changes in legislation, formal decision-making procedures or various types of contractual agreements. Informal institutional change seems to imply changes in norms, values, ideas, traditions, culture, etc. Incremental institutional change seems to imply that changes have built up over time and been released by a decision of some sort or a particular event. Finally, radical institutional change not only implies that the situation, after the radical changes have occurred, is fundamentally different from the situation before but also that the changes come about over a shorter period of time than those which are incremental in nature. In any event, the adjectives used to describe the various types of changes associated with a particular dynamic are largely those also explicitly used by the various theoretical frameworks themselves.

---

[7] See Campbell and Pedersen (2001a, p.12) for a comparable schematic presentation.

Table 2.1: Rational Choice, Historical and Sociological Institutional Perspectives on Institutional Change

Aspects of Institutional Change	Rational Choice Institutionalism	Historical Institutionalism	Sociological Institutionalism
Capturing Institutional Change	• Hypothetical-deductive with methodological individual starting point	• Comparative with a pronounced inductive element	• Hypothetical-deductive but with recent move towards more inductive research strategies
Conditions for Institutional Change	• Shifts in relative prices	• External shocks • Crisis most often seen as objective and non-cognitive phenomenon • Political and economic contradictions • Ideational fit	• Crisis most often seen as ideational phenomenon • Uncertainty about organisational goals, technological causes and effects and external expectations • Window of opportunity
Dynamics of Institutional Change	• Perceived costs and benefits (formal/ informal/ incremental) • Policy entrepreneurs with extraordinary skills and/or enabled by position in institutional environment (none in particular)	• Punctuated equilibrium and feed-back processes (incremental/ radical) • Learning processes (incremental) • Persuasive ideas (incremental/radical)	• Learning processes (informal/incremental) • Diffusion and isomorphism (informal/incremental) • Problem identification, shaping and political selection (incremental) • Policy entrepreneurs with extraordinary skills and/or enabled by position in institutional environment (informal)

In general, the issue of institutional change has not been high on the rational choice institutional research agenda. Yet, it has been suggested that shifts in relative prices are conducive for change and the primary dynamic of institutional change is related to the perceived costs and benefits of alternative institutional arrangements. Rational choice institutionalists also tend to attach an important role to political entrepreneurs in carrying through change. Enabled by their position in a broader institutional environment (either at the margin of the incentives structures of the broader institutional environment or as mediators between constituencies and government officials), and their extraordinary skills (psychological predispositions, knowledge, ability to process information, negotiating skills and persistence) political entrepreneurs may be pivotal in initiating institutional changes. The preferred methodology of rational choice institutionalism is hypothetical-deductive with a point of departure in methodological individualism.

As is the case with rational choice institutionalism, the primary concern of historical institutionalism has not been with institutional change but rather with stability and path-dependency. Yet, over time, a stronger interest in studying and explaining institutional change has also developed within this institutional tradition and a series of suggestions on how to conceptualise the dynamic of ideational and institutional change have appeared. This interest has given rise to the concept of punctuated equilibrium focusing on the interaction between issues and venues, and the conceptualisation of learning processes leading to shifts in the beliefs of the involved agents. Certain ideas are thought to have a persuasive force in themselves, and to the extent that these ideas fit with already existing ideas and institutions they may bring about change. Regardless of the preferred conceptualisations of the dynamics of institutional change, historical institutionalists attach a significant amount of explanatory value to external shocks or change, e.g. general socio-economic disturbances, institutional crisis, change in government, war, etc., in triggering institutional change. The preferred methodology to capture institutional change goes through research strategies with a pronounced inductive element and often takes the form of comparative studies of sector policies or national polities. Common for historical institutional research is a pronounced focus on historical developments and, in particular, a zooming in on periods of time characterised by institution building.

Within the sociological institutionalism, changes are exclusively thought of as incremental and a series of concepts and explanations of this type of institutional change have been put forward. Institutional change may be brought about through learning processes where lessons learned from past experiences may lead to changes in institutionalised routines. Changes may also be generated by processes of diffusion and isomorphism. Such processes involve the spread of norms and ideas between social contexts or across organisational fields. Processes of identification, formulation and the shaping of problems and solutions may produce change and, likewise, the working of policy entrepreneurs may be central to both diffusion processes and in identifying, shaping and bringing problems to the attention of decision makers. Uncertainty within and among organisations, an open window of opportunity and the existence of a crisis are considered conducive to change. The preferred methodology of the sociological institutionalism has been hypothetical-deductive.

3

A Discursive Institutional Approach and its Analytical Implications

The objectives of this chapter are, first, to develop a framework for capturing and understanding institutional change and, second, to construct an analytical strategy that prepares the study of institutional change within the CAP. These objectives are pursued by means of a fourth institutional optic on institutional change – a discursive institutional approach.

A common research agenda is still in the making and institutional research that draws on discourse analysis is still often lumped together with the sociological institutional or social constructivist approaches (e.g. Schneider and Aspinwall 2001). Yet discursive institutional research, which has been on the rise since the early 1990s, has a series of distinctive conceptual and methodological characteristics, which are not easily subsumed, either under the broad sociological institutional or under the more narrow constructivist approach (intimations on this point can be found in Rosamond 2000, p.120ff., 2002, p.364). Against this background, Campbell and Pedersen (2001) recently identified and set out explicitly the contours of a discursive institutionalism, which has the study of institutional change as its central concern. In the following, a number of particularities of a discursive institutional approach to the study of institutional change will be outlined and situated in the context of comparable concepts and conceptions within the rational choice, historical and sociological institutionalisms. It will also be argued how a discursive institutional optic may deal with some of the shortcomings of rational choice, historical and sociological institutional approaches to the study of institutional change: in particular in regard to the type of institutional change which is ideational in nature. In conclusion, a discursive institutional analytical strategy will be drawn up, with the objective of capturing and conceptualising institutional change within the CAP with a particular focus on the articulation and institutionalisation of organic farming within the CAP.

A Discursive Institutional Approach: Conceptualising and Capturing Institutions

The outline of a discursive institutional approach and the conceptualisations of the conditions for, and dynamic of, institutional change draw mainly on Andersen (1995), Pedersen (1995), Andersen *et al.* (1996), Andersen and Kjær (1996), Kjær (1996), Campbell and Pedersen (2001a), Kjær and Pedersen (2001), but inspiration is also drawn from Rochefort and Cobb (1994a), Hajer (1995) and Wittrock and Wagner (1996), and on selected issues even if the latter authors are not explicitly referring to a discursive institutionalism nor do they all have an explicit institutional focus (cf. Rochefort and Cobb 1994a).

Ideas, Discourse and Institutions

The discursive institutional approach to institutional research argued here, takes its point of departure in a logical sequence, which binds together the concepts of idea, discourse, and institution (Andersen 1995; Andersen and Kjær 1996; Kjær 1996; Kjær and Pedersen 2001). Ideas are the final point of reference in which discourses are anchored. Ideas are the final points of reference in the sense that no further explanation is needed or expected when references are made to such ideas. It may be disputed whether or not the articulation of a certain idea is acceptable and legitimate in a particular context, however, the idea referred to is not a matter of dispute. Ideas are the anchor of discourses in the sense that they enable the production of discourse and, for instance, enable the articulation of problems and solutions, while also acting to delimit other problems and solutions from being identified in a particular context (Andersen 1995, pp.18–19). This does not mean that there exists a complete consensus on the articulation of ideas embedded in a given discourse but, rather, that agents need to express themselves for, against and through a set of ideas in order to produce relevant and meaningful statements. In order to speak and act meaningfully and to be taken seriously, agents are expected to refer to a set of commonly recognised ideas. Ideas have in themselves no meaning. However, meaning and possible conflicts over meaning appear when ideas, through processes of articulation, are turned into discourse (Andersen 1995, p.18; Andersen and Kjær 1996, p.8).

Discourses unfold as ideas are articulated and, over time, are turned into rules-based systems of concepts and conceptions and a discourse may thus be defined as 'a system of meaning that orders the production of conceptions and interpretations of the social world in a particular context' (Kjær and Pedersen 2001, p.220). To be able to talk about the existence of a discourse, a system or common set of rules for a collection of concepts and conceptions must be identifiable. Institutions, in turn, are authorised and sanctioned discourse. In general, the set of rules governing a discourse are referred to as institutions when these rules, through processes of institutionalisation, have attained some degree of authority and been linked to certain sanctions (Andersen 1995, p.22). The conceptualisation of institution thus arises out of the concept of discourse, and the relationship between the two is historically contingent and, essentially, an empirical question. More specifically, the institutions identified through a discursive institutional optic are those creating expectations about viable political activity in a particular context by constituting a set of authorised and sanctioned rules on, for instance, acceptable and valid statements, the production and maintenance of knowledge, and the formulation of relevant problems and their solutions (Kjær and Pedersen 2001). The logical sequence between ideas, discourse and institutions, also gives rise to two distinct understandings of change.

Change thus appears: (i) as ideas are turned into discourse, and (ii) as discourse is turned into institutions. The process of ideas being turned into discourse is one of articulation, and the process of discourse being turned into institutions is one of institutionalisation. Since discourses are rules-based systems of concepts and conceptions, processes of articulation progress through the establishment of some sort of discursive rules and, since institutions are authorised and sanctioned discourse, processes of institutionalisation progress through authorisations and the establishment of some sort of sanctions.

The discursive institutional concern with ideational change has similarities with, but also differs from, rational choice, historical and sociological institutional approaches. First, discursive institutionalism, in contrast to recent historical institutional concern with ideas, which are seen as constituting one independent variable among others and exerting a causal effect on the phenomenon selected for investigation (e.g. Hall 1989, 1993), holds that ideas only attain meaning within a particular institutional and discursive context and, as such, should be considered endogenous to the field selected for study. Second, opposed to the historical institutionalism, which tends to operate with ideas as well-defined and stable entities (e.g. Hall 1989, 1993) the discursive institutional approach advocated in the current context does not assume ideational stability. Rather, change in meaning systems or discourses in which ideas are embedded, may also entail changes in the articulation of ideals and, hence, bring about ideational change.

Third, whereas the rational choice and historical institutionalisms tend to address institutions as – albeit humanly constructed – external constraints on human behaviour, the discursive conceptualisation of institutions is closer to the cognitive sociological notion, which holds that agents interpret their environments through institutions. Yet, this discursive institutional conceptualisation of institutions differs also from those of the sociological institutionalism by being more concerned with the rules of 'what is sayable' (Foucault 1991, p.59) than the rules guiding 'what is thinkable'. That is, rather than bringing attention to cultural rules, intentions, and motivations, the rules that are brought into focus by a discursive institutional optic are those that govern the production of discourse and, hence, collective meaning systems. Finally, although institutions are bits of discourse, which have obtained some degree of authority and been linked to sanctions, a discursive conceptualisation of institutions also implies that institutions are not fixed. Rather, institutions are ongoing processes of renewal and possible change and are upheld only for as long as someone actually refers to the authorised and sanctioned discursive rules that constitute institutions (Andersen 1995, p.23).

A Brief Methodological Introduction

Even though studies drawing on discourse analysis – at least in the context of European studies – are often subsumed under the social constructivist label, the discursive institutional approach advocated here differs in particular on methodological issues, in relation to most of its counterparts under the social constructivist label, to a degree that cannot be ignored. More specifically, some practitioners of social constructivist research (e.g. Checkel 2001) inspired by sociological institutionalism, draw on hypothetical-deductive methodologies when studying ideas. A hypothetical-deductive methodology, however, gives rise to epistemological concerns related to reductionism. Particularly in relation to the study of change, it seems that analytically predefined notions of ideas lead to significant limitations as to the range of changes that may, potentially, be captured and subsequently accounted for by such approaches. Conversely, by embedding the articulation of ideas in a given meaning system and making ideational change a central concern for empirical investigation, the discursive institutional approach aims to capture ideational change within the meaning system under investigation irrespective of its nature. Thus, since ideas attain their meaning when articulated and when turned into discourse, *the* distinctive methodological feature of a discursive institutional approach is that it has discourse as its object for study.

The methodological setting of the discursive institutional optic advocated in the current context is neither hypothetical-deductive, as ascribed to by devotees of rational choice institutionalism, as well as some sociological and, to a lesser degree, historical institutionalists – nor is it clear-cut inductive. Rather, the starting point may be described as analytical inductive. On the one hand, the approach is analytical since the point of departure is taken in general reflections on the concepts of, and the logical sequence between, ideas, discourse and institutions as well as in a series of conceptualisations about the conditions for, and dynamics of, institutional change. On the other hand, the approach is inductive in the sense that the analytical frameworks drawn upon in concrete empirical studies are 'drained' of their causalities, concepts are refined and interrelations established through historical descriptive empirical analysis. Essentially, institutional change should be understood through the study of the interrelationship between the discursive and the institutional and this interrelationship is historically contingent.

It is thus important to note that present approach to institutional research has a knowledge ambition that differs from the rational choice and sociological and, to a lesser degree historical institutional approaches, which – to a varying degree – strive to establish causal theories. Against this, rather than striving to achieve the formulation of theories that reflect, and are testable against, an objective, stable and observable reality, the aim is to develop analytical strategies, which in turn enable the study of concrete historical institutional developments (Andersen *et al.* 1996, pp.171–172). In this sense a discursive institutional analytical strategy, on the one hand, takes its lead from theories and concepts of ideas, discourse and institutions while, on the other hand, an analytical strategy obtains its value when it inspires empirical studies, which generate not already available 'insights into aspects of the institutional organization of society' (Andersen *et al.* 1996, p.172).

The Dynamics of Institutional Change: A Discursive Institutional Approach

The discursive institutional approach proposed for the study of institutional change holds that the existence of at least two to each other alternative discourses is a necessary condition for institutional change to appear and, furthermore, that the presence of an ideational crisis is conducive to institutional change. That is, alternatives to the existing discursive and institutional order must exist for this order to be disputed and the existence of an ideational crisis is conducive for the progress of the types of processes of articulation and institutionalisation, which may ultimately come to constitute institutional change. Conflicts over meaning, processes of translation and policy entrepreneurship are all proposed conceptualisations that may capture various dynamics of institutional change. That is, in distinct ways, conflicts over meaning, processes of translation and policy entrepreneurship – either in concert or separately – may give momentum to processes of articulation and institutionalisation, the outcome of which is institutional change. The suggested conceptualisations of the conditions for, and dynamics of, institutional change will be fleshed out below and situated in the context of comparable concepts within the rational choice, historical and sociological institutional approaches.

Alternative Discourse and Conflicts Over Meaning

The discursive institutional approach outlined here holds that the existence of at least two to each other alternative discourses is a necessary condition for institutional change. The reason for this is that it is only in such a situation that a particular institutional context may be contested through disputes over the articulation of the ideas embedded in this institutional context (Campbell and Pedersen 2001a, p.11; Wittrock and Wagner 1996). Apart from constituting a necessary condition for institutional change, conflicts over meaning or conflicts over the articulation of ideas may, however, also be viewed as containing a dynamic, which may give momentum to institutional change.

The rational choice institutional optic on institutional change neither claim nor have a pronounced concern with the formation of preferences outside of what may be deduced from the material conditions of the involved agents, however, both the historical and the sociological institutionalisms operate with conceptualisations of learning processes, which are comparable to the present concern with conflicts over meaning as a dynamic of institutional change. From a historical institutional perspective, the dynamic of change contained in policy-oriented learning processes is often seen as depending on the appearance of technical information, on the rules governing the exchange of viewpoints in a given policy field, and also on conflicts between groupings with different beliefs as to political objectives and causal assumptions (cf. Jenkins-Smiths and Sabatier 1993). It may, however, be argued that this approach assumes too much coherence both of the belief systems between which learning may take place and of the agents ascribing to such belief systems. Hence, this approach leads to analytical blind spots, which renders impossible, for instance, the identification of incoherence and inconsistencies within belief systems and among the agents referring to such. In fact, an important dynamic of institutional change is missed here (I will return to this point below). The sociological institutional conceptualisation of successful learning processes suggests that such processes may be conceptualised as processes of diffusion or isomorphism (cf. DiMaggio and Powell 1991; Strang and Meyer 1994). Although this line of thinking introduces the concept of meaning, the emphasis is still on coherence rather than conflict. Consequently it is unclear whether learning processes are in fact a dynamic of institutional change or, perhaps, instead can be viewed as processes, which increases homogeneity and reinforces already present practises and norms.

Closer to the present concern with meaning conflicts as a dynamic of institutional change is Hajer's (1995) argumentative approach. Hajer suggests that the prime dynamic of change goes through conflicts 'over meaning of physical and social phenomena' (Hajer 1995, p.72) and argues that 'discursive interaction … can create new meanings and new identities' (Hajer 1995, p.59). In order to address the issue of interacting discourses Hajer (1995) makes use of the story-lines concept, which is described as 'narratives on social reality through which elements from many different domains are combined and that provide agents with a set of symbolic references that suggest a common understanding' (Hajer 1995, p.62). Story-lines are thus discursive entities, yet not synonymous with the discursive. Rather, story-lines are seen as metaphors cutting across dissimilar discursive fields, which, in one way or another, all have a share in the construction of a common concern, and contribute to the reduction of the complexity (Hajer 1995, pp.62–68).

The formation of story-lines could then be seen as contributing to change by mediating a shared meaning about a common concern across otherwise dissimilar discursive fields. The main point made by Hajer (1995) is, however, that story-lines may contribute to change as they catch on among a greater number of agents and this may happen to the extent that story-lines appear to be persuasive. A story-line is, in turn, persuasive if it is argued in a credible way, the agents advocating a particular story-line are considered trustworthy and if the practical consequences are acceptable to those agents subject to persuasion (Hajer 1995, p.63). In general, a story-line may gain support to the extent that agents perceive this story to 'sound right' (Hajer 1995, p.61). The notion that ideas must 'sound right' or 'fit' into a particular meaning system in order to be adopted here is, in fact, a recurrent notion in literature addressing the role of ideas in politics (e.g. Hall 1993; Kingdon 1995). Unlike the historical institutional approach to the study of ideas, however, the discursively derived concept of story-lines does not claim that coherency necessarily enforces the spreading of storylines (Hajer 1995, p.61). Likewise, opposed to the historical institutional approach to the study of ideas, the conceptualisation of story-lines does not operate with an analytically predefined notion of the particularity of story-lines which, rather, needs to be identified within a particular discursive and institutional context (Hajer 1995, p.44).

With the objective of capturing the dynamic of institutional change there is, however, also a series of limitations that relate in particular to the concept of story-lines. First, it seems that the introduction of the concept of story-lines overrides the original concern with conflict and instead moves attention towards how story-lines may gain support among an increasing number of agents. In other words, the conceptualisation of story-lines seems to entail a noticeably greater interest with the 'fit' between ideas than with conflicts between dissimilar articulations of ideas. Second, similarly to historical institutional concerns with ideas, those story-lines, which are credible, acceptable and advocated by trustworthy agents, are claimed to have enhanced powers of persuasion. Yet, even though Hajer (1995) seeks to address the question of what 'persuasiveness' may originate from, it seems that the explanatory blind spot is merely passed on and still leaves the question open of the origin of credibility, acceptability and trustworthiness in a particular story-line.

To be sure, the discursive institutional approach advocated in the current context is not in radical opposition to Hajer (1995). In fact, in the name of simplicity the wish is merely to push towards an explication of the link between the institutional and the discursive. A discursive institutional response to the analytical problems outlined is thus that the 'fit' between dissimilar articulations of ideas and the 'powers of persuasion' are to be found in the rules that govern discourse (intimations on this point are also found in Hajer's notions of 'discursive affinities' and 'discursive contamination' (Hajer 1995, pp.66–67). In addition, the present discursive institutional optic wishes to level out the bias towards the emphasis on 'fit' introduced by the concept of story-lines by keeping in mind the issue of conflicts over meaning.

In that sense, the concept of conflict over meaning also aims to level out the bias of previous discursive institutional research towards a focus on 'the history of the winners' or 'histories of hegemonies' (Andersen and Kjær 1996). Even if it is agreed that strategic choices are made during the course of a discursive formation which involves, for instance, certain problems being articulated and institutionalised rather than others, it seems that this concern is rarely turned into an object for discursive institutional empirical research. The concept of conflict of meaning, hence, prepares the field for the study of conflicts over the articulations of ideas and suggests that conflicts over meaning contain a dynamic of institutional change.

Along these lines, the conceptualisation of conflicts over meaning implies, first, that the 'fit' or 'persuasiveness' of ideas as articulated, for instance, in problems and solutions in a given political field, are related to the nature of the rules or institutions governing alternative discourses. To the extent, for instance, an idea finds several non-uniform expressions, yet its articulations are still governed by a set of discursive institutions that have similarities, the way should be paved for such articulations to mutate and possibly for discursive and institutional change. More specifically, for instance, dissimilar problems may notwithstanding be formulated as having similar causes that, in turn, may pave the way for the transfer of solutions.

Second, the discursive institutional perspective suggested, wishes to uphold the conception that disputes over the articulation of ideas contain an important aspect of institutional change and, like above, such conflicts appear around the rules governing discourses. For instance, even though a particular solution may have obtained an institutionalised position in a particular context, conflicts may still evolve around, for instance, the nature of the problem that the solution may resolve. The conflicts, through which a problem is refined, may in turn lead to readjustments in its solution and, hence, the way is paved for discursive and institutional change. Together, it is the 'fit' between and the conflicts over the rules governing discourses that contain the dynamic of institutional change, which is conceptualised as 'conflicts over meaning'.

The meeting between alternative discourses and conflicts over meanings may often leave the institutional context unchanged, yet sometimes this interaction may also lead to mutated meanings, which may give rise to institutional changes over time. That is, first, conflicts over meaning may leave the institutional context unchanged to the extent that the already institutionalised discourse remains intact and its alternatives are rejected. This outcome is referred to as a strategic choice and is the most common outcome since institutionalised discourse, by definition, has the upper hand by already being authorised and linked to some type of sanctions. Second, the outcome of conflicts over meaning may bring about institutional change to the extent that alternative discourse is institutionalised alongside already institutionalised concerns. When conflict over meaning gives momentum to the institutionalisation of alternative discourse alongside already institutionalised concerns, future conflicts over meaning must be expected to appear since conflicts are, in this situation, essentially institutionalised. Finally, the outcome of conflict over meaning may give momentum to institutional change to the extent that a mutation appears out of the meeting of alternative discourses. A mutation is thus the term used to describe the outcome of an interaction between dissimilar discursive rules that have, for instance, produced a combination of problems, their sources and solutions, which differs from the discursive rules as they appeared prior to the interaction and, which through a process of institutionalisation, come to constitute an institutional change. Opposed to the institutionalisation of alternative discourse alongside already institutionalised concerns, when the outcome of conflict over meaning is a mutation of dissimilar discursive rules, then the level of conflict must be expected – at least provisionally – to decrease.

The concept of conflicts over meaning as a dynamic of institutional change has some similarities but also differs from comparable thoughts within the rational choice, historical, and sociological institutional approaches. First, although current study does not aim to specify and conceptualise the processes through which institutional change and change in preferences or interests among agents involved in a given political field are interrelated, it is clear that the two are related. Opposed to, most notably, the rational choice institutional approach, the discursive institutional optic suggested does not, hence, separate preferences or interests from institutions, but rather holds that agent preferences or interests are interpreted through specific historical discursive and institutional contexts.

Second, whereas, for instance, Jenkins-Smiths and Sabatier (1993) assume internal coherence of belief systems and advocacy coalitions, the present approach suggests that discursive incoherence and inconsistency is, in fact, the condition that enables institutional change through conflicts over meaning. That is, mutations of dissimilar discourses appear only when both conflicts exist around the rules governing these discourses – for instance, on the formulation of problems – and where some degree of similarity exists – for instance, on authorised solutions. Moreover, individuals may not be referring to only one particular discourse but are instead more likely to contribute to the production and reproduction of several alternative discourses even where contributions arise out of critical annotations about such alternatives (see also Hajer 1995, pp.69–70). Third, emphasising the duality between 'fit' and conflict as a dynamic of institutional change, prepares the field for the study of mutations of alternative articulations (for instance, of problems) rather than the adoption of cut-and-dried problems and solutions, which is, arguably, the study of continuity as opposed to that of change. Finally, it should be noted that the concept of conflicts over meaning prepares the field for the study of the discursive multiplicity within a particular context and aims to capture a dynamic of institutional change related to the interaction between alternative articulations of ideas within, rather than across, social contexts. This is a central characteristic that distinguishes meaning conflicts as a dynamic of institutional change, as opposed to processes of translation, to which we now turn.

Translation

The second concept proposed to capture a dynamic of institutional change is that of translation. Translation may be described as the 'process whereby concepts and conceptions from different social contexts come into contact with each other and trigger a shift in the existing order of interpretation and action in a particular context' (Kjær and Pedersen 2001, p.219). The concept of translation proposes that agents operating in one social context may select from concepts and conceptions made available to them through contacts with other social contexts. The concepts and conceptions selected may, in turn, be connected to concepts and conceptions already embedded in the context into which they are introduced and, essentially, trigger displacements or change in the existing discursive and institutional order (Kjær and Pedersen 2001, p.219).

    Both the historical and the sociological institutionalisms operate with explanations of institutional change, which are comparable to the conceptualisation of processes of translation proposed from a discursive institutional perspective. Among historical institutional frameworks it is common to refer to external factors or shocks such as broader socio-economic developments, change in government, crisis, or even acts of war as one possible source of institutional change (cf. Hall 1989; Baumgartner and Jones 1993; Sabatier and Jenkins-Smith 1993). From within the sociological institutional approach it has been suggested that a central dynamic of institutional change is to be found in processes of isomorphism or diffusion involving the spreading of standardised practices and models and welldefined and coherent norms within or across organisational fields (cf. DiMaggio and Powell 1991; Strang and Meyer 1994). The concept of translation both has similarities with, and differs from, these historical and sociological conceptualisations of institutional change.

Just as historical institutional research may refer to external factors or shocks as having made an impact on a given policy or polity, or across a number of policies or polities, translation is also a concept that depends, ultimately, on contacts between social contexts. However, opposed to historical institutional approaches that refer to the abrupt impact of external factors or shocks when explaining institutional change, the discursive institutionalism holds that translation – albeit depending on contacts with other social contexts – is an ongoing process among agents within a particular social context (Kjær and Pedersen 2001, p.219). The process of translation has similarities with the sociological institutional conceptualisations of diffusion and isomorphism insofar as these concepts are concerned with the carrying of concepts and conceptions from one context to another. However, whereas diffusion and isomorphism tend to give attention to the spreading of standardised practices and models and welldefined and coherent norms, the conceptualisation of translation focuses on the selectiveness by which certain concepts and conceptions may be elevated from one social context to another. That is, translation is somehow a more complex process than those of diffusion and isomorphism and implies that a selection of certain concepts and conceptions in one social context may – as translation proceeds – displace or mutate with existing interpretations and, thus, trigger a shift in the articulation of ideas embedded in the field in which a translation is taking place. Translation then has to do with the spread of ideas, but ideas are often translated in a selective way and may displace or mutate with already existing articulations of ideas: essentially, the extent to which such displaced or mutated ideas are institutionalised, institutional change may said to have taken place.

It is important to note the difference between the dynamic of institutional change related to the concept of translation, and the dynamic of institutional change contained in the concept of conflicts over meaning, as proposed above. Although both conflicts over meaning and processes of translation are conditioned by the existence of at least two dissimilar articulations of an idea, processes of translation involve – as opposed to conflicts over meaning – contacts with and references to the articulation of ideas within other social contexts. Importantly, this does not make translation an exogenous dynamic of institutional change since the actual processes of change appear as agents within a particular context translate, connect, and displace concepts and conceptions within that context with reference to the articulations of ideas in another context. The concept of translation thus seeks to deal with external factors or shocks as explanations of institutional change, which may degenerate into ad hoc, residual explanations. Finally, whereas the sociological institutional concepts of diffusion and isomorphism are arguably capturing processes of homogenisation rather than change, the concept of translation prepares the field for the study on how concepts and conceptions may be selected in one context, connected to concepts and conceptions in another context, and how this may trigger displacements or mutations and, essentially, lead to change in the existing discursive and institutional order.

From Policy Entrepreneurs to Policy Entrepreneurship

The third and final dynamic of institutional change to be discussed is related to the working of policy entrepreneurs, which is often attributed with a significant function regarding initiating and carrying through change, yet the concept seems often to be under-theorised or lacking consistency. Hence, although policy entrepreneurs are often attributed great explanatory value regarding change in empirical studies, the historical institutional approach is less concerned with theorising over the concept of policy entrepreneurs. Within rational choice institutionalism, it has been suggested that a few key individuals may uphold more unique beliefs and desires and/or positions in the institutional environment, which motivate and enable such individuals to carry out the work of policy entrepreneurs (cf. Downs 1967; North 1990; Downing 1994; Doron and Sened 2001). The explanatory value of policy entrepreneurs in bringing about change seems, however, to depend on a compromise with the basic assumptions about individual rational behaviour and involves attributing certain agents with particular skills and psychological predispositions.

Likewise, sociological institutionalism has suggested that policy entrepreneurs may be central to processes of diffusion as well as in identifying, shaping and bringing problems to the attention of decision-makers. Policy entrepreneurs are considered extraordinarily resourceful and skilful agents whose performance may be enhanced by the institutional environment in which they operate (cf. Kingdon 1995; Checkel 1997). In particular, policy entrepreneurs are most likely to be successful if the ideas advocated by these agents are compatible with those held by political leaders and organisations (cf. Checkel 1997). The sociological institutional approach, however, also tends to attribute policy entrepreneurs with psychological predispositions, which are not easily investigated empirically and often the concept of policy entrepreneurs tends to become a catch-all explanation of change.

There is no immediate discursive institutional response to the objections raised against the rational choice, historical and sociological institutional concern with policy entrepreneurs, and discursive studies do not often leave much room for the study of political agency. Within discursive institutional research there seems to exist a reluctance to include the study and qualification of agent positions or roles and the focus is usually exclusively on twists and turns in discursive formations and processes of institutionalisation. As far as can be seen there is, however, no reason why we should not pursue the study of, and seek to qualify, how various agent positions within discursive and institutional environments may be constructed and the consequences of such constructions. In fact, Wittrock and Wagner (1996) suggest that:

[a]llowing a focus on spatiotemporal specificities that provide the precise conditions under which social actors are able to give meaning to their existence in the world and to formulate strategies for action … is a necessary turn away from a view on history as the actorless evolution of abstract, universal processes. To realise the full potential of such a conceptual reorientation, it is essential to consider institutional structures not merely as constraints on human action, but as sets of historically established rules of action and as 'containers' of resources and rules, which actors can draw upon and which enable them to pursue their courses of action (Wittrock and Wagner 1996, p.107).

Although this is a call for the study of the role of particular agent positions, it should be clear that agents are not attributed particular non-contextual and exogenously given characteristics. Rather, it implies a focus on individual or collective agency as discursively and institutionally constructed positions. Closer to the issue of policy entrepreneurship, this is also the point made when Hajer (1995, p.51) claims that '[t]he influence of a stubbornly resisting actor…cannot be explained by reference to the importance of his position alone, but has to be given in terms of the rules inherent in the discursive practice, since they constitute the legitimacy of his position'. Hence, in order to move beyond ad hoc, residual explanations of policy entrepreneurs guided by 'inner directions' or the 'objective' importance of a position held by a certain agent, the discursive institutional approach advocated in the current context takes its starting point from the suggestion that policy entrepreneurship is a position enabled by the rules governing a particular discourse.

It is proposed that policy entrepreneurship may be defined as a political role upheld by individuals or collective agents and from where momentum is given to processes of articulation and institutionalisation, which, in turn, bring about discursive and institutional change. In addition, a typology of policy entrepreneurship may be established which distinguishes between translators, creators of forums for communication and carriers.[8] First, the concept of policy entrepreneurship prepares the field for the study of the agency, which contributes to processes of translation. This is considered the more vigorous type of policy entrepreneurship since it is exercised through the linking of concepts and conceptions in one context, to concepts and conceptions in another context, and this may trigger discursive displacements or mutations, which – to the extent such displacements or mutations are institutionalised – come to constitute institutional change. Second, policy entrepreneurship may also be exercised in relation to the establishment of a 'meeting place' or forum for

---

[8] The two first mentioned types of entrepreneurship of the three-fold typology proposed draw some inspiration from Sverrisson (1999) who suggests that the position of entrepreneurs may be offering the holder the opportunity of 'networking' (bringing people together), the opportunity of producing 'knowledge' and interpretations and/or the opportunity of introducing 'innovations'.

communication or negotiations. This type of policy entrepreneurship prepares the field for the study of the agency, which contributes to the establishment of a meeting place that brings together agents and enables the production of meanings. That is, this type of policy entrepreneurship may contribute to processes of articulation and institutionalisation and, hence, institutional change by enabling the authorisation of certain concepts and conceptions, and the creation of related sanctions among the involved agents. Third, the concept of policy entrepreneurship as proposed here prepares the field for the identification of those individuals or collective agents that have contributed to the carrying of those concepts and conceptions institutionalised over time which, in turn, come to constitute an institutional change in a given context. The latter is considered the less vigorous type of policy entrepreneurship. On the one hand, this type of policy entrepreneurship contributes to the institutionalisation of certain concepts and conceptions through authorisations and possibly the establishment of sanctions. Yet, on the other hand, this type of policy entrepreneurship involves reiterations of concepts and conceptions in a particular context rather than the linking of concepts and conceptions from one context to another, as it is the case with policy entrepreneurship contributing to processes of translation.

It is important to note that a single individual or collective agent is unlikely to bring about changes in a discursive and institutional order. Yet, discursive or institutional change is given momentum by someone – not any one or anybody – and it is the agency, which gives momentum to processes of articulation and institutionalisation and, essentially, institutional change that the concept of policy entrepreneurship aims to capture. Moreover, the position of policy entrepreneurship as suggested here is a position different individual or collective agents may take up at different points in time and they may exercise policy entrepreneurship individually, simultaneously or successively. Finally, the current study does not so much aim to illuminate the discursive and institutional construction of a particular agent position (see Pedersen (1988) for such a study) but rather, the ambition is to identify the agency, which have given momentum to articulations and processes of institutionalisation that over time come to constitute institutional change.

The suggested discursive institutional approach to the study of policy entrepreneurship as a dynamic of institutional change has some similarities with, but also differs from, thoughts on policy entrepreneurs within the rational choice, historical and sociological institutional approaches. First, discursive institutionalism differs from the historical institutional approach to the working of policy entrepreneurs by suggesting that we study policy entrepreneurship as a political position or role with certain qualities, rather than retreating to ad hoc explanations involving named individuals or agents. Second, similar to the rational choice and sociological institutional approaches, the discursive institutional concept of policy entrepreneurship suggests that the institutional environment enables this position or role and, hence, the exercise of policy entrepreneurship. Third, however, opposed to the rational choice and sociological institutional approaches, the discursive institutional approach proposed in regard to the study of policy entrepreneurship does not attribute policy entrepreneurs with particular skills and psychological predispositions but, rather, holds that policy entrepreneurship is a position or role enabled only by the discursive and institutional context.

Crisis as an Ideational Condition for Institutional Change

As already noted, the discursive institutional approach holds that at least two to each other alternative discourses must exist for institutional change to come about, but also that the existence of an ideational crisis is conducive for institutional change (Campbell and Pedersen 2001a, p.11). The latter is a point shared by the historical institutional approach where crisis is often seen as an objective and non-cognitive condition for institutional change, triggered by external events (cf. Hall 1989; Baumgartner and Jones 1993; Sabatier and Jenkins-Smith 1993). Most recently, however, historical institutional research, inspired by discourse analysis, has proposed to study crisis as an exclusively ideational phenomenon (Hay 2001). Along these lines, crisis is thought to appear through the identification of contentious issues associated with the dominant discourse and a general 'perception of the need to make a decisive intervention' within the political field at hand (Hay 2001, p.193, 203). Moreover, rather than conceive of crisis in terms of revolutionary points in time, a crisis is seen as evolving over time as a 'crisis narrative' unfolds (Hay 2001, p.202).

The discursive institutional approach advocated here is in concord with this latter view, insofar as crisis is considered conducive for institutional change in the sense that it may create a space of possibility for alternative ideas, for instance, through alternative formulations of the problems to deal with and solutions to be implemented. The discursive institutional optic proposed is also in concord with the above, inasmuch as crisis should be studied as an ideational phenomenon that unfolds over time. Yet, the discursive institutional optic – as given in the present context [9] – also differs from Hay's approach on predominantly methodological issues.

First, rather than studying crisis through 'process-tracing' or 'thick historical descriptions', the discursive institutional study of crisis as an ideational phenomenon goes through discourse analysis. That is, it is through a study of the

---

[9] Hay (2001) is included in a section on discursive institutional takes on 'The rise of neoliberalism' in Campbell and Pedersen (2001). Hay himself, however, refers to his approach as 'ideational institutionalism', which particularly on methodological issues is more closely associated with recent historical institutional research than with the discursive institutionalism as the latter is advocated by Kjær and Pedersen (2001) and in present context.

articulation of ideas regarding crisis that the existence of a crisis may be identified. Second, rather than establishing general models of the causal relationships between crisis, paradigm shift, and policy change (Hay 2001, p.201), the discursive institutional approach holds that articulations regarding crisis must be analysed as part of a particular discursive and institutional context. Conceptions of what makes up an crisis and, importantly, any possible articulated and institutionalised responses are, thus, considered to be embodied in particular discursive and institutional contexts and, hence, not readily comparable across contexts.

It has been suggested that policy fields considered highly contentious and overloaded with issues conceived of as problematic, may be readily characterised as being subject to a crisis (Rochefort and Cobb 1994a, pp.21–22). However, the discursive institutional approach proposed here holds that a crisis exists when it is conceived to exist in a given discursive and institutional context and, in this regard, most importantly, when such conceptions are widely expressed across the context. That is whereas non-uniform articulations of problematic issues may easily be identified within a particular political field, an ideational crisis is more extensive in the sense that it involves – if not unanimous agreement – a high degree of concord of its existence among the involved agents. Finally, it is important to note that whereas the existence of an ideational crisis is conducive to processes of articulation and institutionalisation that may, ultimately, come to constitute institutional change, it does not imply that the involved agents necessarily make explicit institutional choices based on rational calculations as demanded by rational choice institutionalism.

Defining the Terms 'Linguistic Field' and 'Policy Field'

So far, terms such as social context and policy field have been used without supplementary explanation. However, before summing up on the conditions for, and dynamics of, institutional change along the lines of the discursive institutional approach proposed, it is pivotal to point out what is meant by the term field since all of the proposed concepts in some way or another depend on the ability to distinguish between fields. Kjær (1996, p.18) offers a definition of a social field and distinguishes such a field from a linguistic field. Accordingly:

[t]he process of institutionalization can...be seen as a process that results in the construction of a social field, a stable network of institutionally defined positions from which it is possible to act and articulate, that are separated from other fields by a common border and related to each other through certain mechanisms'. Moreover: '[w]hile discursive practice constitutes a linguistic field i.e. symbolic universe in which further articulations can take place, the institutionalization of discursive distinctions constitutes a social field, i.e. a social universe in the context of which social practices and social relations are interpreted and regulated.

It is helpful to distinguish between a linguistic field and an institutionally constructed field and, essentially, to distinguish between fields that are the product of processes of articulation and institutionalisation respectively. However, the term policy field will be preferred henceforth to characterise an institutionally constructed field. Therefore, along the lines of Andersen (1995, p.41), I suggest that a policy field can be said to exist insofar as it is possible to identify a system within which disputes evolve around a given common concern among a set of agents through commonly recognised processes and, to the extent that it is possible to distinguish such a field of concerns, agents and processes from other fields.

Summary: The Dynamics of Institutional Change

The discursive institutional conceptualisations on the conditions for, and dynamics of, institutional change as proposed above, is summarised in Table 3.1. First, it is proposed that the existing discursive and institutional order needs to be disputed by the presence of at least one alternative discourse in order for institutional change to appear.

Table 3.1: Conditions for and Dynamics of Institutional Change

Aspects of Institutional Change	Conceptualisations	Features
Conditions for Institutional Change	• Ideational crisis • Alternative discourse	• Widespread conceptions of crisis • Contestations of the existing institutional order
Dynamics of Institutional Change	• Conflicts over meaning • Translation • Policy entrepreneurship	• Conflicts and 'fit' between articulations of ideas • Selections, displacements, mutations • Translators, creators of forums for communication, carriers

Second, it is suggested that widespread conceptions of a crisis within a particular discursive and institutional order is conducive to processes of articulation and institutionalisation, the outcome of which is institutional change. Third, apart from being a necessary condition for change, the interaction between alternative discourses is proposed to contain a dynamic of institutional change related to, on the one hand, conflicts over meaning or the articulation of ideas and, on the other hand, the 'fit' between alternative articulations of ideas.

Fourth, it is proposed that processes of translation contain a dynamic of institutional change that recognises the selectiveness by which agents in one context may adopt concepts and conceptions from other contexts, connect these with concepts and conceptions within the existing meaning system and, hence, trigger displacements or mutations that ultimately may bring about institutional change. Processes of translation within a given discursive and institutional context thus depend on contacts with other contexts. Finally, policy entrepreneurship is suggested as a political role from which momentum is given to processes of articulation and institutionalisation, which in turn bring about institutional change and – when exercised – may involve contributions to processes of translation, the establishment of forums for communication and the carrying of particular concepts and conceptions.

The conditions for, and dynamics of, institutional change proposed above interact in as yet unspecified ways. The existence of alternative discourse is seen as a necessary condition and conflicts over meaning may possibly give momentum to processes of articulation and institutionalisation that come to constitute institutional change. Yet, conceptions of the existence of a crisis in a given field may also be related to the existence of an alternative discourse – as it may highlight alternative articulations on what should be considered as problematic issues and legitimate solutions. Translations are the processes through which ideas are selected and translated in a particular context, yet such processes may also be related to the exercise of policy entrepreneurship. Policy entrepreneurship is thought of as a subject position from where momentum is given to processes of articulations and institutionalisation, the outcome of which is institutional change, but it is also a position enabled by the institutional expectations attached to that position in a given field. Finally, widespread conceptions of the existence of a crisis are thought of as conducive to institutional change but may also be a reflection of severe meaning conflicts. These are merely a few indications of the possible interrelationships between the conditions for, and dynamics of, institutional change that are at work.

A Discursive Institutional Analytical Strategy

Previously in this chapter, a brief introduction was given to the methodology of the discursive institutional approach. Building on those thoughts and on the conceptual reflections made above, the following will present an analytical strategy for the study of institutional change within the CAP: in particularly the type of institutional change which is ideational in nature. First, it will be discussed how problems and solutions may be studied as ideational symptoms. Second, it will be argued how the articulation and institutionalisation of organic farming within the CAP may be illustrative of the possible institutional changes taking place within the CAP and, hence, illustrative of the usefulness of the conceptualisations of the conditions for, and dynamics of, institutional change proposed by the discursive institutional approach. This part will be concluded by an outline of measurements of institutional change and an overview of, and reflections upon, the empirical material consulted in the subsequent empirical analysis.

Problems and Solutions as Ideational Symptoms

A distinctive feature of the discursive institutional approach to the study of institutional change is that change is captured through descriptive analyses of particular discursive and institutional developments. In the words of Foucault '[t]he important thing is to give the monotonous and empty concept of 'change' a content, that of the play of specified modifications' (Foucault 1991, p.58). In that sense, the first step in a discursive institutional analytical strategy is to capture the articulation and institutionalisation of ideas in a particular discursive and institutional context.

Doing so enables us to say something about how ideas are turned into discourse and something about how discourse is turned into institutions and, essentially, prepare the groundwork for us to say something about the conditions for, and dynamics of, institutional change. The articulation of problems and solutions are often considered particularly illustrative of the ideas embedded in a discourse or meaning system (Hajer 1995; Pedersen 1995; Kjær and Pedersen 2001). The way problems find their expressions are seen as linguistic symptoms of a set of ideas, which can be understood only within a particular discourse and set of institutions (Pedersen 1995; see also Hajer 1995) and, when studied over time, ruptures in the formulation of problems may be used as estimates of possible ideational change (Pedersen 1995; Kjær 1996, p.22). Dependent on the degree to which particular problem formulations have been authorised and possibly linked to sanctions, a process of institutionalisation may also be said to have taken place. The study of problem formulations and their authorisation is, in other words, the study of symptoms of the articulation and institutionalisation of ideas. The formulation of a problem may further be seen as a speech act, which set out the conditions, for instance, for the types of solutions that are applicable and legitimate in a given situation (Kjær and Pedersen 2001). The institutions in focus are thus those rules which, over time, come to govern – albeit not determine – how problems may be formulated, the rules that enable the identification of the sources of such problems, and those rules that guide the involved in their search for legitimate solutions. We still need, however, to address the issues of: (a) how a problem may be encircled; (b) when is it possible to characterise a problem as a discursive phenomenon; and, essentially, (c) the issue of when a problem may be said to have been institutionalised.

As to the issue of how a problem may be encircled, Pedersen (1995) suggests that we should be looking for the 'fundamental categories' in the language used in a certain policy field in order to describe the 'anatomy' of the problems at hand. Along these lines, it is suggested that within a given discursive context problems may be instituted theoretically, normatively, analytically and cognitively. Theoretically instituted problems may appear, when theoretical causal assumptions are coupled with empirical data and when the insight of this coupling in turn establishes a concern about future development. Normatively instituted problems may appear when predicted developments are coupled with ideals about unwished or wished for developments and when the insight of this coupling in turn establishes a concern about future developments and, possibly, a need to make a choice. Analytically instituted problems may appear when predicted developments are coupled with actual developments and when the insights of this coupling are turned into some sort of experience about the relation between the predicted and the actual development. Finally, cognitively instituted problems may appear when individuals or collective agents couple theoretical causal assumptions with empirical data, couple predicted developments with ideal developments, as well as evaluate predicted against actual developments (Pedersen 1995, p.19). While is may be difficult to categorise particular problems in a clear-cut way, the above may be seen as typology of ways that problems may appear within a discursive and institutional context; yet it lacks specific instructions as to how problems may be encircled empirically.

Although Rochefort and Cobb's (1994) social constructivist take on processes of problem definition is not explicitly drawing on an institutional framework, their reflections on the nature of problems may be helpful and seem to fit into the discursive institutional optic argued in current context. Rochefort and Cobb (1994), hence, emphasise the importance of problem formulations for political change. It is also assumed that the way problems are formulated among agents involved in a given field will guide political activity in this field. Moreover, the study of change, in terms of how political agents formulate what they have identified as central problems within a given political field, is expected to tell us something about what type of political activity may be viable. That is, as argued from a discursive institutional perspective, the expression of a political problem is in itself seen as a political activity, which sets out a space of possibility for, among other things, the sorts of solutions that may be adopted.

Against this background, Rochefort and Cobb (1994a) introduce a series of categories and distinctions concerned with the causalities surrounding problems, and applicable and legitimate solutions (see Table 3.2). Rochefort and Cobb (1994a, pp.15–17) suggest a distinction between individual and impersonal or structural causal explanations as helpful when considering problem causality. It should thus be considered whether the emergence of a problem is accounted for by referring to, for instance, either human faults and incompetence or, alternatively, the failure of previous policies or unhelpful social structures. It is expected that political initiatives are most likely to be taken when the causes of a problem are expressed as being of a structural nature or caused by the failure of previous policies. Another distinction may be found between simple or multi-causal explanations. If a problem is described by a simple causality, it may indicate that political action is timely as otherwise, if a problem is formulated as being of multi-causal origin, it may indicate that the problem must be reconsidered before it is possible (and considered effective) to take political action.

Table 3.2: A Problem Definition Perspective

Categories	Distinctions	Suggested causality
Problem Causality – What are the causes of a problem?	• Individual/structural • Simple/multi-causal • Episodic/thematic	• Structural causes are more likely to generate public regulation • Simple causes suggest political action is timely/ multi-causal origin has greatest chance of mobilising support over time • Episodic causes indicate one-off incident/ thematic causes is most likely to generate pressure for change
Solutions – What are legitimate and applicable solutions?	• Availability • Acceptability • Affordability	• Political action depends on the availability of a solution • The likeliness of political action increases when a solution is in accordance with prevailing ideas • The likeliness of political action increases when adequate resources to address the problem are perceived to exist

A third distinction to make when studying the nature of problems has to do with whether the causes of a problem are thematic or episodic. When causes are expressed as being thematic rather than episodic, political pressure for change is more likely to accumulate. Finally, political initiatives will also depend on whether a solution has been formulated, that a potential solution is in accordance with prevailing ideas, and that the resources exist to deal with the problem at hand (Rochefort and Cobb 1994a, pp.24–26).

It should be noted that the discursive institutional approach advocated in current context does not assume any pre-established causality between the institutional rules governing the definition of problems at any given point in time and subsequent institutional change. Also, rather than dissociating the framing of problems from ideas, it is important to note that the discursive institutional concern with problem formulations is essentially reflecting the notion that problems are linguistic symptoms of a particular set of ideas. Still, Rochefort and Cobb (1994a) outline a set of categories and distinctions to encircle problems, which are useful in the sense that they prepare the way for identifying the sources of the problems of concern, the causations involved in the definition of these problems, and the available, acceptable and affordable solutions in a particular context. In addition, with the point of departure in Table

3.2 and the discursive institutional line of thinking outlined, it is also possible to address the issues of when problems as ideational symptoms have been turned into rule-based discourse and when problems have been institutionalised.

In general terms, a problem can be said to have been turned into a discursive phenomenon when is it is possible to describe a set of common rules for the articulation of this problem. That is, it is possible to identify a set of rules for the cause(s) of a problem along the lines of the distinctions individual/structural, simple/multi-causal, and episodic/thematic as well as a set of rules for the availability, acceptability and affordability of the solution(s) that may resolve the problem at hand. Likewise, a problem may be said to have gone through a process of institutionalisation, when it has been authorised and tied up to sanctions. That is, the extent to which the problem at hand has attained an institutionalised status is determined by: whether a problem has been authorised as an issue that needs to be addressed, whether expectations about political action have been created, and where breaching such authorisations and expectations are linked to sanctions. For instance, problems that have been addressed by public legislation which, if breached, will release some kind of sanction, are considered to have a high degree of institutionalisation. Problems may also be endorsed by statements of political intentions through, for instance, various types of policy papers, action plans or white papers. Sanctions may be legally instituted but may also be socially instituted. For instance, problems lacking readily available, acceptable and/or affordable solutions may be brushed aside as irrelevant, frivolous or as having dubious scientific foundations.

Altogether, the discursive institutional approach to the study of problems as a study of the articulation of ideas and processes of institutionalisation has some similarities with, but also differs on several accounts from, the rational choice, historical, and sociological institutional approaches. First, whereas some rational choice institutionalists (cf. North 1990) give marginal attention to ideas in the study of institutional change, the discursive institutional concern with the study of problems reflects the central interest in the articulation of ideas of this approach.

Second, opposed to historical institutional approaches concerned with the definition and redefinition of issues in the study of institutional change (cf. Baumgartner and Jones 1993), the discursive institutional optic links the study of problems to an understanding of ideas, which does not assume ideational coherence and nor does it operate with analytically a priori defined ideas. Rather, the study of the expressions that problems find in a given discursive and institutional context, is the study of the particular articulations that ideas find in this context whatever their nature may be and however diverse and incoherent they may appear. Third, and contrary to sociological institutional concern with the identification, shaping and selection of problems for political consideration (cf. Kingdon 1995), the discursive institutional perspective outlined explicitly links the focus on the formulation of problems with processes of institutionalisation and, hence, institutional change. Finally, by drawing on social constructivists on the framing of problems (cf. Rochefort and Cobb 1994) some analytical instructions as to how problems may be encircled have been introduced into a discursive institutional optic.

Encircling the Empirical Research Object

It has already been argued that the CAP constitutes a critical case for the study of institutional change by commonly being considered to make up a policy field highly resistant to change. It has also been argued that institutional change may be studied as the outcome of processes of articulation and institutionalisation and, thus, that change come about over time. Further, it has been argued that whereas the existence of alternative discourse is a necessary condition, the existence of an ideational crisis is conducive for institutional change to appear. Finally, is has been argued that processes of articulation and institutionalisation and, thus, institutional change may be given momentum by conflicts over meaning, processes of translation and policy entrepreneurship.

Against this background, an issue has been selected for further investigation, namely, the articulation and institutionalisation of organic farming within the CAP. To be sure, although organic farming as an agricultural practice has been subject to growth in the EU it is still a niche in terms of the land area used for organic farming and in terms of the of number farms practicing organic farming compared to the total agricultural sector. For instance, in 1985, the land used for organic farming was estimated to make up less than 0.1% of the total land available for agricultural production in the EU. In 2001, this figure had increased to 3.3%. The number of organic farms of the total number of farms in the EU is estimated to have increased from 0.37% in 1993 to 2.3% in 2001 (Commission 2002, p.9; Foster and Lampkin 1999, p.16). These figures, however, cover great variations in terms of the size of organic farming sectors within Member States. Accordingly, of the total area available for agricultural production, in 2001, it was estimated that the land used for organic farming was around 8% and 11% in Austria and Sweden respectively, between 6% and 8% in Denmark, Finland and Italy, around 3.5 – 4% in the UK and Germany and, in the remaining EU Member States, less than 2% (Organic Centre Wales 2003).

In qualitative terms, or more accurate, in terms of the processes of institutionalisation, however, the relationships between organic farming and the CAP can be expected to involve changes, which reflects developments beyond relevance for the organic farming sector. That is, such processes can be expected to involve developments in conceptions related to, for instance, the relation between, on the one hand, agriculture and, on the other hand, the environment, the use of technology and food quality (compare to, for instance, the possible institutional construction of a policy field or a regime concerned with the regulation of potato production within the CAP). Accordingly, the articulation and institutionalisation of organic farming within the CAP is a research object that appears to have a number of character traits, which may make it illustrative for the study of institutional change. First, on the one hand, organic farming as an agricultural production method and as a range of ideas about agricultural production is often traced back to early in the 20th century (see, for instance, Padel 2001, p.42). On the other hand, in legal terms organic farming has been a separate agricultural sector and an object for Community regulation under the auspices of the CAP since the early 1990s (Lampkin *et al.* 1999). Organic farming has, in other words, a history outside of the CAP but also history inside the CAP and, thus, qualifies as an issue, which over time has moved from not being an object for Community regulation to being one under the auspices of the CAP. Second, it also appears that there exist a dual 'fit' and 'conflict' between ideas articulated regarding organic farming and the CAP. By way of introduction, this may be illustrated by comparing two 'manifestos', namely, the objectives of the EU agricultural policy as they were drawn up under the Treaty of Rome and 'The principle aims of organic production and processing' as formulated by The International Federation of Organic Agriculture Movements (IFOAM). The original Article 39 of the Treaty of Rome from 1958 is still in place (although after the Treaty of Amsterdam it was renamed Article 33) and states that the objectives of the Common Agricultural Policy are:

(a) to increase agriculture productivity by development of technical progress and by ensuring the rational development of agricultural production and the optimum utilisation of the factor of production, particular labour;
(b) to ensure thereby a fair standard of living for agricultural populations by the increasing of the individual earnings of persons engaged in agriculture;
(c) to stabilise markets;
(d) to guarantee regular supplies; and
(e) to ensure reasonable prices in supplies to consumers (Minet 1962).

IFOAM, on the other hand, among other things, states that organic farming should:

produce food of high quality in sufficient quantity…[and]…consider the wider social and ecological impact of the organic production and processing system…[and]…use, as far as possible, renewable resources in locally organised production systems…[and]…minimise all forms of pollution'. Organic farming should also: 'allow everyone involved in organic production and processing a quality of life which meets their basic needs and allows an adequate return and satisfaction from their work, including a safe working environment…[and]…progress toward an entire production, processing and distribution chain which is both socially just and ecologically responsible (IFOAM 2000).

On the one hand, there exist an apparent 'fit' between the two manifestoes, for instance, as to the concern with people in agriculture, which should be able to uphold a decent standard or quality of life, to receive a fair return in terms of earnings for their work and to supply sufficient quantities of food products. On the other hand, there also exist at least potential conflicts over the emphasis to be put on, for instance, the objective to increase agricultural productivity by means of technological progress vs. the objective to consider the social and ecological impact of agricultural production by means of renewable resources. Furthermore, potential conflicts may exist between the two manifestos on the emphasis to be put on, for instance, the objective to supply of food products at reasonable prices and the objective to produce of high quality food products. The adoption of organic farming as an object for Community regulation may thus be expected to involve either certain changes in the way the CAP objectives are conceived of among the involved agents, or certain changes in the conception of organic farming as articulated by the international organic farming movement. Alternatively, it may be that certain changes have appeared in conceptions related to both the CAP objectives and ideas articulated regarding organic farming. Whichever type of institutional changes that may be observed empirically, it seems that they may potentially be illustrative of the usefulness of the conceptualisation of conflicts over meaning proposed as a dynamic of institutional change by the current study.

Third, as mentioned, organic farming appears to have moved from constituting a set of ideas and an agricultural practice outside of the auspices of the CAP to become an object for Community regulation within the CAP. This process, or such processes, may potentially be illustrative of the usefulness of the conceptualisation of processes of translation, which is proposed as a dynamic of institutional change that is conditioned by contacts between different social contexts. Fourth, someone have given momentum to the articulation and institutionalisation of organic farming within the CAP, and, arguably, this 'someone' have been enabled by a discursive and institutional context, which is not immediately in complete keeping with the objectives of the CAP as adopted by the Treaty of Rome.

Hence, the exercise of policy entrepreneurship regarding the articulation and institutionalisation of organic farming within the CAP may be expected to involve certain discursive and institutional changes. Further, arguably, the exercise of policy entrepreneurship giving momentum to the articulation and institutionalisation of organic farming within the CAP may also be expected, at least at certain points in time, to have been in conflict with the prevailing discursive and institutional context. In other words, the focus on the articulation and institutionalisation of organic farming within the CAP may potentially be illustrative of the usefulness of the concept of policy entrepreneurship as a dynamic of institutional change as proposed by the current study.

Fifth, organic farming has been proposed to constitute a social movement or something resembling an alternative discourse to that of conventional agriculture (Campbell and Liepins 2001; Kaltoft 2001; Michelsen 2001). Although, organic farming may not in a strict sense constitute a completely distinguishable discourse, it appears that organic farming finds its nuance against conventional agriculture and represents a set of values, objectives and practices which, to some extent, have been defined in opposition to those of conventional agriculture. In that sense, organic farming appear to represent a set of alternatively articulated ideas to those – at least for certain periods of time – prevailing within the CAP, which the current study consider a condition for institutional change. Finally, it has already appeared from the CAP reform literature that the CAP at certain points in time has witnessed a crisis regarding budget pressures. The current study, however, approach the study of crisis as an ideational phenomenon and include, for instance, crisis regarding food safety issues that have been witnessed within the CAP and, which may potentially have been conducive to the articulation and institutionalisation of organic farming within this field.

Altogether, the current study approach the CAP as a critical case for the study of institutional change, and focus on the articulation and institutionalisation of organic farming within the CAP as reflecting processes of change beyond relevance for organic farming and as potentially illustrative for the usefulness of the discursive institutional approach proposed above to the study of institutional change: in particular the type of institutional change which is ideational in nature.

The current study is also delimited in time so to focus on the time span from 1968 to 2005. As mentioned, organic farming was formally adopted as a sector for political regulation under the auspices of the CAP in the early 1990s. However, it is possible to trace links between organic farming and Community policy objectives back to the early 1970s and, likewise, it is possible to identify developments in the relationship between organic farming and the CAP after the early 1990s. Moreover, by focussing on the time from 1968 to 2005, the study will expand over all of the formally appointed CAP reforms, that is, the reforms in 1968, 1984, 1988, 1992, 1999 and 2003, and the most recent initiative to elaborate an EU-wide action plan for organic food and farming will also be included.

Measurements of Institutional Change Within the CAP

The point of departure of current the study is a logical sequence between ideas, discourse and institutions. Change is thus considered to appear: (i) as ideas are turned into discourse, and (ii) as discourse is turned into institutions. Changes related to ideas being turned into discourse, or rules-based systems of concepts and conceptions, is the outcome of processes of articulation, which progress through the establishment of some type of discursive rules. Change related to discourse being turned into institutions, or authorised and sanctioned discourse, is the outcome of processes of institutionalisation, which progress through authorisations and the establishment of some sort of sanctions. It has also been argued that problems may be studied as symptoms of ideas embedded in a particular discourse. Against this background, how do we measure institutional change empirically?

Essentially, measurements are needed for when a problem – as an ideational symptom – has been turned into rule-based discourse and for when a problem has been institutionalised. The current study will consider that a problem has been turned into rule-based discourse within the CAP to the extent it is possible to identify a set of common discursive rules for the articulation of this problem among concerned agents within the EP, the Commission or the Commission services, and the Council. Such discursive rules have to do with the causality of the problem on hand, which may be described along the lines of the distinctions individual/structural, simple/multi-causal and episodic/thematic problems causes. Likewise, a solutions has been turned into rule-based discourse within the CAP to the extent it is possible to identify a set of common discursive rules for its availability, acceptability and affordability among concerned agents within the EP, the Commission or the Commission services, and the Council.

The current study considers that a problem or a solution has been institutionalised within the CAP when the discursive rules governing the articulation of the problem or solution on hand have been authorised and tied up to sanctions. Problems and solutions articulated in *legal texts* such as directives and regulations adopted by the Council are considered to represent the highest degree of institutionalisation. The reason for this is that such texts have usually gone through protracted formal and informal processes of selection and authorisation among a wide range of agents and are, if infringed, often linked to various kinds of formal as well as informal sanctions. Problems and solutions articulated in *Commission Green/White papers and other Commission Communications* are considered to represent a certain degree of institutionalisation since the elaboration of such documents involves a wide range of agents, which refer to these documents as authoritative even if infringements are dealt with by sanctions that are more informal. That is, agents that do not express themselves for, against, or through the problematic issues and solutions raised by these documents, may not be taken seriously in the debate.

More specifically, *Commission Green papers* are publications which will usually contain a series of formulations of what are considered to be problematic issues in a given policy area as well as suggestions on how these problems could be addressed. Green papers then form the basis for further discussion among a selected group of interested agents with the objective of specifying the exact nature of the problems at hand and possibly to set out strategies for courses of action. On the other hand, *Commission White papers*, which are usually preceded by a Green paper, and *other Commission Communications* contain the specific problems to be dealt with and proposals for legislation in a particular field. In spite of the differences in scope and specificity, any Commission Green/White paper and Communication on the CAP tends to generate attention and debate. Plans drawn up with the explicit intention to reform seem particularly helpful when we are to study ideas and their institutionalisation since it is fair to expect that it is when reforms are in the pipeline that problems – as ideational symptoms – find their way to the forefront and their clearest expressions.

Problems and solutions articulated in *EP resolutions, reports, debates and questions to the Commission* are not necessarily representing a high degree of institutionalisation within the CAP. Rather – if nothing else is known – problems and solutions articulated in EP resolutions, reports, debates and questions to the Commission are taken to represent a certain degree of institutionalisation within the EP depending on the extent to which the various texts are endorsed by the members of the EP. Likewise, problems and solutions articulated in *policy papers, conference speeches, Member State reports and Parliamentary debates* are – if nothing else is known – taken to represent a degree of institutionalisation within the organisation, administrative unit, Member State or party, which have published the document in question.

Articles from *Agra Europe* (and few other newspaper articles) have also been consulted. *Agra Europe* is an independent publication, which has followed CAP developments in detail since 1963, and is distributed weekly among farmers' organisations, Agriculture Ministries and other policymakers close to the CAP. Rather than taking the problems and solutions articulated here as representing a degree of institutionalisation within, for instance, the media, these texts are consulted in order to identify the ideas drawn upon by a broad range of involved agents and, importantly, to get an unbroken and constant source of information. *Agra Europe* is to some extent used as a 'seismograph' of articulations on the subject matter particularly – but not exclusively – among EU Member States and interest organisations. This is done since information about Member States' contribution to the production of discourse within the Council is otherwise difficult to obtain. Importantly, after having consulted both primary documents – for instance national agriculture committee reports – and the secondary reporting of such documents in *Agra Europe*, it is clear that the essential formulations appearing in the reports referred to, very closely resemble their original wording. Moreover, the articles appearing in *Agra Europe* often contain longer quotations from the referred to reports or speeches. It should, however, be noted the editorial style of *Agra Europe* tend to identify with a market-oriented and agri-business perspective on the Common Agricultural Policy.

In summary, the articulation of problems and solutions – as ideational symptoms – in legal texts and Commission Green and White papers are taken to represent the highest degrees of institutionalisation within the CAP. Ideas articulated in the other types of texts consulted are taken to represent a degree of institutionalisation among those agents endorsing and authorising the texts, and possibly representing the endorsement of alternative articulations of ideas to those enjoying an institutionalised status within the CAP. It is possible in that way to identify how certain articulations of ideas may be elevated from less to more authoritative texts and, thus, identify institutional change. Moreover, by consulting texts such as speeches by Commissioners and Commission bureaucrats, speeches by Ministers for Agriculture and the Environment, policy papers and *Agra Europe* articles about links between the CAP and organic farming in the EU, it is possible to identify alternative and conflicting ideas articulated among the involved agents.

The empirical measurements thus aim to, first, identify processes of articulations and institutionalisation and, hence, discursive and institutional change respectively. Second, the aim is to identify alternatively articulated ideas and, hence, possible conflicts over meaning and, third and related, to identify possibly processes of translation to the extent that ideas articulated within the CAP are enable by their contacts with other related fields. It is important to note that it is obviously only possible to cover a certain range of fields, which may have made ideas available for translation within the CAP. It, hence, seems particularly relevant to have a view to, for instance, the environmental policy field and alternative farming communities and otherwise to be guided by references in the documents consulted. Fourth, the focus on the agents endorsing particular concepts and conceptions aim to identify possible policy entrepreneurship exercised regarding the articulation and institutionalisation of organic farming within the CAP. In this regard, it will also be noted whether certain agents have contributed to the establishments of forums for communication such as, for instance, hearings and conferences. Finally, it will be noted whether the involved agents refer to the existence of a crisis when voicing the need to deal with a particular problem or the implementation of a certain solution within the CAP.

An Overview of the Empirical Material

In the present study a wide range of different types of documents will thus be used, namely, Commission White and Green papers and other Communications; EP resolutions, reports, debates and questions to the Commission; legal texts such as directives and regulations; policy papers; conference speeches; Committee reports; Member State reports and articles from *Agra Europe*. Table 3.3 is an overview of the number of documents used per source and type of documents. All of the types of documents used are obtained in their electronic version where possible, in order to enable electronic searches across a large volume of empirical material. Official EU documents (Green/White papers, EP reports, debates and parliamentary questions, legal texts) dating back to the early 1990s have, to a large extent, been available electronically and some even further back. Likewise, some policy papers and conferences speeches have largely been available electronically back to the mid-1990s.

Table 3.3: Overview of the Empirical Material

Sources	Total number of documents per source	Numbers of documents per type of document
Agra Europe (and other newspaper articles)	N=123	Agra Europe n=116 Others n=7
Commission and Commission Service	N=37	Reports n=20; Speeches n=9; Answers to EP questions and press releases n=8
European Parliament	N=48	Reports and resolutions n=6; Speeches n=35; Questions n=7
Council	N=16	Legislations and declarations n=9; Press releases n=7
Member State reports and speeches	N=29	DK n=5; NL n=4; AT n=2; DE n=3; EL n=1; SE n=3; UK n=11[10]
Organised Interests	N=11	COPA n=2; Organised organic farming interests n=7; EEB n=2
Total	N=264	

Wherever possible, full text documents were obtained of the Member State reports referred to by *Agra Europe* and academic literature. In particular, the empirical material use to capture ideas articulated in relation to organic farming in the 1970s and 1980s is scarce and marginal to the CAP overall, however, some of the available material from this particular resource is, in fact, significant regarding the development of organic farming. That is, documents that can be regarded as somewhat marginal in terms of the CAP in general, are central to the evolving of organic farming in order to capture the developments that first take place at the periphery of the CAP. Even so, later, it will be seen that these apparently marginal developments at the periphery, come to constitute actual changes under the auspices of the CAP. Finally, four exclusively explorative interviews have been conducted with centrally placed EU and UK bureaucrats as well as with representatives of COPA and the Soil Association, which is the largest UK based organisation representing organic farming interests.

---

[10] This figure includes two reports concerning organic farming developments in Wales and one report concerning organic farming developments in Scotland.

4

The Translation and Institutionalisation of the World Problematique (1968–1977)

This chapter aims to establish the pretext for subsequent articulations and institutionalisations linking organic farming to the CAP. Along these lines, the 1968 Mansholt Plan will be consulted in order to capture the ideal conceptions articulated within the CAP at this point and enable the identification of subsequent changes within the CAP. *'The Limits to Growth – A Report for the Club of Rome's Project on the Predicament of Mankind'* was published in the early 1970s, and Hajer (1995, pp.78–84) has identified it as something of a breakthrough for environmental politics across the western world. In fact, *The Limits to Growth* became an international bestseller and the message of an increasingly present 'world problematique' caused by the almost intuitively true notion of the world being finite was received as pointing to a series of legitimate problems among world leaders (Hajer 1995; see also Lomborg 1998). That is, *The Limits to Growth* will be drawn upon in order to capture the ideal conceptions articulated by means of the 'world proplematique'. The problems identified by *The Limits to Growth* are also recognised as legitimate within the European Community and, in fact, the first Community Environmental Action

Plan from 1973 explicitly refers to it as a reason for the emerging Community concern with environmental issues. The first and second Community Environmental Action Plans will be consulted in order to investigate the implications of the ideals made available by *The Limits to Growth* within the emerging EC environmental policy from 1973 to 1977.

It will be proposed that the ideal conceptions articulated and made available in the context of *The Limits to Growth* by means of the 'world proplematique' are selectively translated within the emerging EC environmental policy. The ideal conceptions selectively translated within the emerging EC environmental policy in turn form the basis for the articulation and institutionalisation of a relationship between agricultural production and environmental depletion, as well as the articulation of biological farming as something of increasing interest among farmers and consumers. Altogether it appears that ideas articulated within the context of *The Limits to Growth* and the emerging EC environmental policy forms the basis for subsequent processes of articulation and institutionalisation and, hence, institutional change within the CAP.

The Mansholt Plan (the CAP – late 1960s/early 1970s)

The first EU Commissioner for Agriculture (1958–72), Sicco Mansholt, had his name given to the first attempt to reform the CAP in 1968 – being known as 'The Mansholt Plan'. The background for the initiative to reform the CAP was a period characterised by vast economic expansion and income increases outside of the agricultural sector, as well as a rising agricultural production that exceeded Community consumption levels and contemporaneous exports demand (Kommission 1968, pp.1–7).

Two central problems are identified in this regard. One of these problems may be characterised as normatively instituted in the sense that it is based on a coupling between an actual, but also a predicted, further deepening of existing societal welfare disparities that disfavoured farmers and an ideal conception holding that public policies ought to balance out such disparities. That is, one problem for the CAP is how to deal with an agricultural community, which is lagging behind the welfare boost enjoyed by the rest of society. The other problem may be characterised as analytically instituted in the sense that it is based on a coupling between, on the one hand, predicted developments in agricultural markets and, on the other hand, the insights obtained on the consequences of previous price policies and certain structural characteristics of the agricultural sector. That is, the other central problem of the CAP is one of mounting output and the future prospect of surplus production across an increasing number of agricultural products and, hence, disturbed agricultural markets and rising agricultural expenditures (Kommission 1968, pp.1–16). The sources of the mounting problems within agriculture are to be found in the current Community price policy as well as in the structure of the agricultural sector (Kommission 1968, p.16).

Against the background of rising productivity in the agricultural sector, in combination with slower growth in agricultural consumption and export markets, the current high price policy is increasingly putting strains on the expenditures of the Community budget to the agricultural sector (Kommission 1968, pp.4–5). Furthermore, the problem of an agricultural community not benefiting from general societal welfare expansions are attributed to structural developments within the agricultural sector itself, such as an ageing farmers' population, low education levels, low adaptability, for example, in terms of technological progress, as well as the existence of too numerous, so-called, 'borderfarms', which are medium-sized farms unable to sustain a reasonable quality of living for a farm family (Kommission 1968, pp.5–6, 20, 26–27).

Whereas the 'technological and industrial revolution' carries at least some of the responsibility for the rising economic welfare of society at large (Kommission 1968, p.17), technological progress, both contributes to certain problems, and is a fundamental condition for the future of European agriculture. On the one hand, '[a] high price policy and progress within chemistry, animal health, plant protection and genetics have led to considerable growth in outputs' (Kommission 1968, p.31; author's translation from German). That is, technological innovations are contributing to increases in surplus production and its associated problems. At the same time, the march of technological progress cannot be stopped and the structure of the agricultural sector needs to adapt to industrial and technical developments in order to be on a more equal footing to share the welfare benefits enjoyed by society at large (Kommission 1968, p.20, 28). Apart from the duality of technological progress (i.e. it contributes to surplus production but is also a 'fact of life'), the causes of the problems identified by the Mansholt Plan are clear and simple, and structural and thematic in nature (high price policy, structural features of the agricultural sector, technological progress) and found both inside (high price policy) and outside the CAP (structural features of the agricultural sector, technological progress as a 'fact of life').

The objective of the Mansholt Plan is two-fold, namely, to improve the balance of agricultural markets and to improve a number of structural aspects of the agricultural sector (Kommission 1968, p.8). The solutions envisaged are expressed in a section of the Plan entitled 'Agriculture 1980 Program'. From this programme it appears that the current market and price policy is not sufficient to deal with the agricultural problems of the Community, which will become even more severe if action is not taken immediately, as well as being accompanied by the implementation of a series of measures concerning the medium term. In addition to a restrictive price policy, the measures to be taken are aimed at significantly reducing the number of farms and employees in agriculture, to increase farm size, to increase transparency in agricultural markets and enhance production discipline, that is, sufficient, consistent and homogeneous supplies (Kommission 1968, pp.29–39). In general, the Mansholt Plan has the explicit objective of modernising the agricultural sector and improving its competitiveness (Kommission 1968, pp.1–3).

As it turned out the Mansholt Plan was not adopted in its original form but rather three watered-down 'socio-structural directives' were agreed upon in 1972. These directives involved provision of financial support to farmers for improving farm efficiency, support to farmers who wished to take early retirement and make their land available for the enlargement of other farms, training for farmers who wished to stay in the agricultural sector, and re-education for farmers wishing to find other jobs outside the agricultural sector (Ackrill 2000, p.211). The most common explanation for the lack of success of the Mansholt Plan is that it was considered far too radical in both Germany and France. In Germany, the Ministry of Agriculture very early on brushed aside the plan as being too costly and Mansholt was even branded the 'peasant killer' by the West German Minister for Agriculture (Josef Ertl) (Fennel 1997, pp.218– 219; *Agra Europe* 19.10.1979). The central problems identified by the Mansholt Plan were also, seemingly, in radical opposition to the concerns of French agricultural policy at the time. Hence, in France:

the debate took place against a background of concern about the prospects for the farm community, faced by acute problems of transitions from a peasant-type, small-scale farming pattern, depending almost exclusively on family labour and making little use of purchased inputs, to larger-scale, mechanised and capital-intensive units (Tracy 1989 quoted in Grant 1997, p.71).

Additionally, Member State governments were, in general, reluctant to enforce a policy with the explicit ambition of considerably reducing the number of people working in agriculture while it was highly uncertain that this surfeit of manpower could find alternative employment outside the agricultural sector (Ackrill 2000, p.211). Finally, it has been suggested that the Mansholt Plan basically failed due to its advocacy of a type of 'modern farm', which was in stark contrast to a core principle of domestic agriculture that emphasised 'the centrality of small-scale farming (chiefly in the southern Member States) and family farming (predominantly in northern Member States) to the (re)structuring of rural space' (Jones and Clark 2001, p.23).

The main point in the present context is that, at this point in time, articulations of, for instance, a relationship between agricultural production and the environment or, for instance, of a link between alternative agricultural production methods and certain benefits, does not appear within the context of the first attempt to reform the CAP. Rather, such links are, partly, established outside of the Community in the context of *The Limits to Growth* and, partly, within the emerging EC environmental policy.

The Identification of the World Problematique (early 1970s)

'The Club of Rome', which is made up of approximately 70 scientists, policy-makers and leaders from the private sector, was established around the same time that the Mansholt Plan first saw the light of the day. It was this group, which in 1972 published *The Limits to Growth – A Report for the Club of Rome's Project on the Predicament of Mankind*. Above all *The Limits to Growth* identifies a 'world problematique' involving 'poverty in the midst of plenty; degradation of the environment; loss of faith in institutions; uncontrolled urban spread; insecurity of employment; alienation of youth; rejection of traditional values; and inflation and other monetary and economic disruptions' (Meadows *et al.* 1972, p.10). The 'world problematique' may be characterised as normatively instituted in the sense that it is based on a coupling between predicted exponential growth on a number of factors and an ideal conception holding that there are limits to growth and the need to strive towards a 'global equilibrium'.

The claim is that exponential growth in the world population, agricultural and industrial production, the consumption of non-renewable natural resources and pollution will eventually lead to a collapse – very possibly within the next hundred years. The collapse will take the form of the depletion of non-renewable resources, malnutrition, an abrupt fall in world populations, and industrial capacity, and will be outside the control of man. Although technological innovations may be able to address some of these problems in the short term, new technology and, for example, the discovery of additional stocks of non-renewable resources will only postpone, not eliminate, the inevitable collapse (Meadows *et al.* 1972, chapter 4). The ideal conception is simple: the world is finite and there is an absolute limit to growth. Acknowledging that, the interrelationships between the above-mentioned five exponential growing elements have set the world on track towards disaster and radical change is needed.

While a 'world problematique' has been identified, the *The Limits to Growth* offers only ambiguous and general instructions as to how policymakers may respond to this problematic. The basic claim of *The Limits to Growth* is that the 'world problematique' may only be understood – and ultimately addressed – holistically through an approach that considers the interplay between 'technical, social, economic and political elements' (Meadows *et al.* 1972, p.11). Policymakers have to give up the idea that growth is possible in the long term and instead steer towards 'a state of global equilibrium', which implies that a balance between populations and capital must be established within the limits of the world material resources. The striving for 'a state of global equilibrium' is presented, however, as depending on fundamental cognitive changes. *The Limits to Growth*, hence, emphasises that 'any deliberate attempt to reach a rational and enduring state of equilibrium by planned measures, rather than by change or catastrophe, must ultimately be founded on a basic change of values and goals at individual, national, and world levels' (Meadows *et al.* 1972, p.195; see also p.9). Although the point of departure is taken in the assumption that the 'world problematique' may only be understood as a complex interplay between economic, political, social and technical factors, what comes out at the other end is a simple mono-causality.

Simply put: growth (in population, agricultural and industrial production) is not possible in the long term since growth depends on the consumption of non-renewable resources. Additionally, the causes of the 'world problematique' are clearly structural and thematic in nature (depletion of non-renewable resources, malnutrition, abrupt falls in world populations and industrial capacity outside the control of man). Finally, in order for the 'world problematique' equation to add up, it is assumed that technological progress does not provide – in the long term – a solution to the exponential growth in the consumption of non-renewable natural resources. Rather than supplying solutions, technological progress tends to contribute to the formation of the 'world problematique'. That is, technological innovations in industrial production, for example, in the use of agri-chemicals and fertilisers in intensive agricultural production, contribute to pollution, depletion of non-renewable resources and malnutrition and, hence, push the world towards collapse (Meadows *et al.* 1972, pp.51–54; chapter 4). As is appears, some of the ideal conceptions articulated in the context of *The Limits to Growth* also appear within the emerging EC environmental policy.

The Emerging EC Environmental Policy (1973–1977)

Environmental policy was not made a treaty obligation until the implementation of the Single European Act in 1987, yet Community concerns with the environment is commonly traced back to a decision by Heads of State and Government in 1973, which recognised the protection of the environment as an objective of the Community (e.g. Sbragia 2000, p.296). In November 1973, the Council adopted the first of so far six Environmental Action Plans, which sets out the environmental objectives of the Community. A wide range of agents was heard and involved in the development of the first Environmental Action Plan, including Member State governments, the Commission, the EP, the Economic and Social Committee and labour market organisations (Weale *et al.* 2000, p.56). Neither the first Environmental Action Plan nor its successors are legally binding texts, yet they are based on wide political endorsements and, hence, may be seen as attached with some degree of authoritativeness and establishing certain expectations about future action.

Against this background, the first Environmental Action Programme, for instance, claims that '[t]he natural environment has only limited resources; it can only absorb pollution and neutralize its harmful effects to a limited extent. It represents an asset which can be used, but not abused, and which should be managed in the best possible way' (Council 1973; Title II Principles of a Community Environment Policy). This type of formulation resembles the normatively instituted 'world problematique' as it is articulated in the context of *The Limits to Growth* and although the analytical lines drawn between the articulations of certain ideal conceptions are not conditioned by explicated references, the first Environmental Action Programme does, in fact, explicitly refer to *The Limits to Growth* (Council 1973; Reasons for action and guidelines). Yet it is also clear that greater hope is attached to further technological progress within the emerging EC environmental policy than is offered in the 'world problematique' equation. For instance, the emerging EC environmental policy states that:

[t]he best environment policy consists in preventing the creation of pollution or nuisances at source, rather than subsequently trying to counteract their effects. To this end, technical progress must be conceived and devised so as to take into account the concern for protection of the environment and for the improvement of the quality of life at the lowest cost to the Community' (Council 1973; Title II Principles of a Community Environment Policy). In addition: [t]he standard of scientific and technological knowledge in the Community should be improved with a view to taking effective action to conserve and improve the environment and to combat pollution and nuisances. Research in this field should therefore be encouraged' (Council 1973; Title II Principles of a Community Environment Policy).

It is unclear whether it is the diverging time perspective of *The Limits to Growth* and the emerging EC environmental policy that leads, respectively, to distrust and trust in technological progress. Nevertheless, it seems that the trust attached to technological progress is critical for the availability, acceptability and affordability of solutions within the emerging EC environmental policy. In other words, it seems that a trust in technological progress rules out solutions in accordance with a 'global equilibrium' as preferred in the context of the *The Limits to Growth'*, and instead prepares the search for other types of solutions.

The emerging EC environmental policy also embraces the notion that the protection of the natural environment, in a significant way depends on agricultural policies. Two specific problems related to the agricultural sector are identified. These problems have to do with the depopulation of certain rural regions in the Community, and the affect of agriculture on the natural environment, and both of these problems are articulated as being related to environmental depletion. On the one hand, the depopulation of agricultural land is problematic since farmers are considered to have a significant role in taking care of the countryside. On the other hand, the affect of agricultural production is related to 'the intensive use of certain types of fertilizer and the misuse of pesticides' and with the aim to protect the environment 'the dangerous effects of such practices should be lessened' (Council 1973; Title II Action to Improve the Environment). Although it is recognised that the CAP has contributed to the modernisation of agriculture, scientific knowledge is rather limited on the ecological effects of modern production techniques and, hence, research in this area should be initiated (Council 1973; Title II Action to Improve the Environment). Links are thus established between the CAP and the modernisation of European agriculture as well as between the modernisation of agriculture and environmental depletion. Yet, the latter link is still weak and lacks a scientific basis and, hence, is only articulated by suspicion. A direct link between the principles of operation of the CAP and environmental depletion is yet to be established

Finally, a very early authoritative expressed concern with organic or biological farming within the context of the Community appears in the context of the emerging EC environmental policy. The first Environmental Action Plan thus holds that 'consumers are increasingly paying attention to the quality of foodstuffs' and 'farmers are also increasingly developing "biological" products or products obtained by methods which are "closer to natural processes". Further research is needed, however, in order to investigate the authenticity of "natural" and "biological" products both in relation to production techniques and marketing' (Council 1973; Title II Action to Improve the Environment). That is, consumers and farmers are gradually displaying an awareness of 'biological' food products and production but it is still doubtful as to whether these products represent a real solution to fulfilling consumer concerns. To be sure, the articulation of a link between organic farming (or, biological farming, which is the commonly used phrase at this point in time) and agricultural policy concerns appears within the still emerging EC environmental policy – not within the CAP. Moreover, it should be noted that the articulation of a relationship between biological farming and EC agricultural policy concerns are scarce and marginal to the overall objectives of the emerging EC environmental policy. While a link between biological farming and agricultural concerns can not be characterised as institutionalised within the emerging EC environmental policy since further research is still needed on the matter, it represent an early articulated relationship between biological or organic farming and Community agricultural policy objectives.

References are no longer made to organic farming or food products, yet the problems and their causes identified in relation to agriculture in the first Environmental Action Plan are very much carried through unaltered into its successor, which came into force in 1977. It is still the depopulation of certain regions and the environmental affects of agriculture on the environment, which are given attention (Council 1977; Non-damaging use and Rational Management of Land; Introduction). If anything, even greater trust is placed in technological progress as a means of addressing a wide range of environmental problems (Council 1977). It is still emphasised that farmers have a role in protecting the environment and it is clarified that modern agricultural production techniques do in fact have adverse effects on the environment (Council 1977; Measures relating to rural areas and forestry). That is, by 1977, the source of the environmental problems in agriculture is simple, structural and thematic in nature, and found outside the CAP (modernisation of agriculture). As with the previous Environmental Action Plan, potential solutions to environmental problems are contained in the 'activities' of 'the farmer', yet 'modern production techniques' produce environmental problems (Council 1977; Measures relating to rural areas and forestry).

Translations and Institutionalisations (1968–1977)

The problems and solutions as they are articulated within the CAP, in the context of the 1968 Mansholt Plan, within the context of *The Limits to Growth* and within the emerging EC environmental policy, may be summed up as shown in Table 4.1. In the late 1960s/early 1970s, the central concern within the CAP, as identified by the Mansholt Plan, is agricultural surplus production and an agricultural sector lagging behind industrialisation processes and income developments outside of agriculture. The sources of these problems are the high price policy, certain structural aspects of the agricultural sector and, hereunder, a lack of adaptability of certain farms. The solutions envisaged have to do with a CAP in support of the modernisation of agriculture, measures to improve the competitiveness of the agricultural sector, and a restrictive price policy.

In the early 1970s, an alternative conception of the current socio-economic developments appears in the context of *The Limits to Growth* which, in a broad sense, gives rise to conceptual changes among political leaders across the western world. The ideal conception is here that modernisation and industrialisation push the world towards disaster. That is, exponential growth in populations, in agricultural and industrial production, in the consumption of non-renewable resources and in pollution is leading to a global collapse. Technological progress is not going to reverse this development since there is an absolute limit to the potential of growth. Rather, technological innovations contribute to the 'world problematique' since modern technology tends to enhance industrial production, the consumption of non-renewable resources and pollution. A balance between populations and capital must be established within the limits of the world material resources – a global equilibrium.

Although the latter notion is lost in the process, the 'world problematigue' as made available in the context of *The Limits to Growth* is translated within the emerging Community environmental policy where it forms the basis for the articulation and institutionalisation of environmental problems caused by agricultural production. That is, processes of institutionalisation and, hence, institutional change within emerging EC environmental policy is conditioned by alternative ideas made available in the context of *The Limits to Growth* and is given momentum through a process of translation.

Table 4.1: Problems and Solutions within and around the CAP (1968–1977)

	The CAP (late 1960s/early 1970s)	The World Problematique (early 1970s)	The Emerging EC Environmental policy (1973–1977)
Problems	• Agriculture lagging behind societal welfare increases • Surplus production	• World problematique	• Rural exodus in certain regions • Effects of agriculture on the environment
Causes	• High price policy • Structural features of agricultural sector • Lack of adaptability in terms of technological progress	• Exponential growth in population, agricultural and industrial production, consumption of non-renewable resources and pollution • Technological progress	• Modernisation of agriculture
Type of Problem and Causality	• Normatively/analytically instituted problems • Mono-causality • Structural and thematic causes • Combination of causes inside and outside of the CAP	• Normatively instituted problem • Mono-causality • Structural and thematic causes	• Normatively instituted problems • Mono-causality • Structural and thematic causes
Solutions	• Modernisation of agriculture • Improvement of competitiveness • Restrictive price policy • Technological progress as 'a fact of life'	• Global equilibrium • Technological solutions not viable in the long term	• Technological progress
Organic Farming as a Solution	• None	• None	• Consumers increasingly interested in quality food products including biological food products (acceptability) • Farmers increasingly interested in natural production methods including biological production (acceptability) • Research needed to clarify authenticity and benefits of biological products (not readily available)

By 1977, agricultural problems relating to the rural exodus in certain regions of the Community and, in particular, the effect of agriculture on the environment, have been institutionalised within the emerging EC environmental policy. The reason for both developments is found in the modernisation of the agricultural sector, which involves more intensive agricultural production methods and an increasing use of agri-chemicals. In order to address these problems Community regulation is needed and technological progress is here a key to the generation of solutions. Moreover, it is recognised that consumers are increasingly showing an interest in biological products as food products of a particular quality and, likewise, farmers are showing an increased interest in biological farming as a natural production method. However, the real benefits of such products and production methods are still to be determined by further research into the matter.

Finally, whereas it is central to the equation of the 'world problematique' as formulated in the context of *The Limits to Growth*, the translation of the 'world problematique' within the emerging EC environmental policy proceeds selectively, and the conception that technological progress does not contain viable solutions in the long term is not adopted in this context. That is, solutions in accordance with a 'global equilibrium' are not selected for translation within the emerging EC environmental policy but, rather, there is the idea endorsed and institutionalised that further technological developments should be encouraged in order to address problems of environmental depletion. In that sense, by 1977 the 'world problematique' had been subject to a mutation and institutionalised accordingly in the context of the emerging Community environmental policy, which entailed that the resolving of environmental problems should be pursued through technological innovations.

5

The Translation and Institutionalisation of Environmental Ideas within the CAP (1978–1985)

The introduction of environmental concerns into the CAP is often traced back to 1985 and the adoption of a Council Regulation (797/85), which among other things allowed EU Member States to support environmentally friendly farming and take national measures to protect particularly environmentally sensitive areas (Council 1985). It has been suggested that the UK government was especially eager to push forward this regulation and that 'little favour' was initially found among other Member State governments as well as within the Commission (Lowe and Whitby 1997, p.294; Lenshow 1998). Yet the Council Regulation was passed since the implementation of the vast majority of the measures included was optional for Member States (Fennel 1997, pp.230–231). Even so, the following will propose that regulatory initiatives such as the above were conditioned by the existence of alternative ideal conceptions of agricultural problems and solutions and, prior to 1985, enabled by the translation of such conceptions within the CAP. Likewise, it is proposed that the advocacy or policy entrepreneurship in favour of such initiatives was conditioned by contacts with alternative ideal conceptions of agricultural problems and solutions, and enabled by their translation within the CAP prior to 1985. It is also proposed that widespread conceptions of the existence of a crisis in the first half of the 1980s seem to have been conducive to certain processes of institutionalisation. Essentially, it is proposed that the period from 1980 to 1985 is characterised by the translation of an ideal conception, which links intensive agricultural production to certain environmental problems and enables, among other things, the articulation of organic farming as a potential and partial solution to some of these problems in the context of the CAP. It is these discursive and institutional developments, which will be considered by consulting a total number of 18 empirical documents and articles.

The Institutionalisation of Environmental Ideas within the CAP (1980–1985)

As early as in 1975 the Commission had adopted a communication elaborated by the DG for Agriculture, which stated that 'agriculture can…have certain unfavourable effects on the natural environment. In particular efforts should be made to mitigate the dangerous consequences of certain modern production techniques' (Commission 1975 as quoted in Jones and Clark 2001, p.28). The conceptions that modernised agricultural production may have adverse effects on the environment resemble the conception that was first articulated, and from which any lingering doubts were later removed, and accordingly institutionalised within the EC environmental policy by 1977. However, when this conception was voiced within the CAP in the mid-1970s it was abortive. It has been argued that the link between modern agriculture and certain environmental problems was rejected at the time because it fitted very poorly into the idea holding that environmental concerns are essentially setting up constraints on the agricultural sector, which were a conception occupied by Member State agriculture ministries (Jones and Clark 2001, p.29).

Against this background, the problems central to the CAP in the beginning of the period from 1980 to 1985 may be characterised as cognitively instituted in the sense that they are very much a reflection of those problems first predicted and feared in the context of the 1968 Mansholt Plan. However, by the end of the period a series of normatively instituted problems had been institutionalised alongside the then still in force cognitively instituted problems. The problems institutionalised by the end of the period may be characterised as normatively instituted in the sense that they are based on a coupling between, on the one hand, an actual but also a predicted further modernisation of agricultural production, increasingly severe budgetary constraints, and depressed markets and, on the other hand, an ideal conception holding that the CAP should contribute to the protection of the environment. Although organic farming is neither institutionalised in any legal sense, nor can it be seen as anything resembling a policy field: organic farming is articulated and endorsed as a potential and partial solution to certain problems within the CAP and is a case in point of conceptual developments during the period in question.

Agricultural Ideas Within the Commission (Early 1980s)

In late 1980 the Commission adopted a communication entitled 'Reflections on the Common Agriculture Policy' elaborated by the DG Agriculture. As it appeared, the prime problems facing the CAP were the continuing increase in agricultural production, which 'engenders an uncontrollable rise in expenditures' (Commission 1981, p.8). Surplus production and rising agricultural expenditures, in turn, were seen to be caused by the use of the regulatory mechanisms favoured by the CAP, which entailed farmers being supported through guaranteed prices and direct product subsidies on unlimited product quantities (Commission 1981, pp.8–9). Rising agricultural expenditures had to be curbed, but it was considered particularly problematic that the larger and more production effective farmers and already fortunate regions were favoured by the CAP support mechanisms then in place. Echoing the ideas given voice by the Mansholt Plan, the problem in this regard was not seen as developments towards larger and more efficient productions units but rather the use of public funding to support the already better off farmers that gave rise to concern (Commission 1981, pp.8–9).

While the support mechanisms operated in favour of the large and rational farmers, the problem of supporting farmers already working under the more fortunate natural and structural conditions was essentially related to developments outside of agriculture. Hence, it was the protracted economic recession, along with an energy crisis as well as an impending Community enlargement to the South, which gave rise to concerns about the use of public funding (Commission 1981, p.9, 13). On the one hand, the functioning of the support regime of the CAP gave rise to problems related to an unjust distribution of public funds. On the other hand, the concerns related to the use of public funds were essentially brought about by developments considered external to the CAP and related to the energy crisis, the then still present economic recession, the natural and structural condition of agriculture and a forthcoming enlargement of the Community.

It is worth noting that the CAP contributed to the modernisation of European agriculture and, hence, the optimisation of production factors and the increasing productivity (Commission 1981, p.7). However, this development in agriculture was not immediately linked to problems of surplus production. Rather, by encouraging the modernisation of the agricultural sector during the 1960s and 1970s, the CAP contributed to 'the remarkable boom in the industrial and tertiary sectors by providing them with the necessary labour' (Commission 1981, p.7). Moreover, technological developments were conceived of as a 'fact of life' and as possibly containing solutions to problems that related to energy consumption in agriculture, rather than singled out as a source of the current problems related to surplus production. For instance, on the one hand, it was argued that 'in the present state of agricultural technology it is neither economically sound nor financially feasible to guarantee price or aid levels for unlimited [product] quantities' (Commission 1981, p.12). On the other hand, the agricultural sector 'has an urgent need for technologies' that may minimise the substantial consumption of energy in this sector (Commission 1981, p.16). Opposed to the ideals of the emerging EC environmental policy, technological solutions are, however, by no means the preferred course of action within the CAP. Although without specifications as to the exact nature of the actions to be taken an emphasis was, rather, put on the potential of further exports of agricultural products to markets external to the Community as well as the setting up of restrictions on imports of products, which were already being produced within the Community. It is considered 'unjustifiable to criticize the operation of the CAP while leaving the door completely open to competing products for political or other reasons' (Commission 1981, p.15). In addition to this protectionist strategy, it was considered that the socio-structural aspects of the CAP should be enhanced including increased support to 'less-favoured areas', the modernisation as well as cessation of certain farms and afforestations. However, the greatest hope was attached to the concept of co-responsibility. It was considered that the support of farmers through guaranteed prices and direct product subsidies leading to agricultural production independent of market demands should be counteracted by the introduction of co-responsibility on behalf on the farmers. Co-responsibility, basically, entailed that, above a certain level of production, the otherwise guaranteed prices for agricultural products should be reduced (Commission 1981, pp.13–14).

The problems as read by the Commission in the beginning of the 1980s are cognitively instituted in the sense that they are based on two dissimilar couplings. Accordingly, the problems are based on a coupling between evaluations of the developments predicted as early on as the late 1960s and actual developments over the past 15 years (Commission 1981, pp.20–34). At the same time, the problems are also based on a coupling between, on the one hand, an actual and further predicted developments disfavouring certain farmers and agricultural regions in the Community and, on the other hand, an ideal holding that the CAP should strive towards balancing out such disparities. The causes of the problems articulated are multiple, clearly structural and thematic in nature (guaranteed prices and direct product subsidies, economic recession, energy crisis, natural and structural conditions of agriculture, forthcoming Community enlargement) and found both inside the CAP (guaranteed prices and direct product subsidies) and outside the CAP (economic recession, energy crisis, Community enlargement). Finally, acceptable and available solutions were held to be: a protectionist CAP, the modernisation of certain regions as well as the introduction of a co-responsibility on behalf of the farmers (Commission 1981, p.15, 17).

Organic Farming and Policy Entrepreneurship Within the EP
(Early 1980s)

A report drawn up by the EP Committee on Regional Policy and Regional Planning (hereafter the EP Committee on Regional Policy) and a motion for a resolution endorsed by the EP (Europa Parlamentet 1982) confirms, in late 1981, a number of the problems of European agriculture that had been advocated by the Commission a year earlier. That is, agricultural surplus production and the pressure of agricultural expenditures on the Community budget were clearly still regarded as critical issues. It was confirmed that, although the CAP had a positive effect on the upholding of reasonable incomes and the lifestyle of farmers, it was favouring some farmers and regions over others: the protracted economic crisis was also cited as a source of the problems being experienced (European Parliament 1981, pp.5–6). Echoing the concept of co-responsibility, formulated by the Commission a year earlier, solutions envisaged by the EP involved the setting up of limitations on the quantities of production to receive guaranteed prices and the establishment of a link between financial support and either the amount of farm land under cultivation or the number of livestock, rather than production output (European Parliament 1981, p.6).

As it appears, some of the problems, their causes and solutions advocated in the EP report fit in with those identified by the Commission in late 1980 and, hence, represent continuation. Yet, additional concerns were voiced about problems relating to the degradation of rural life and rural exodus, the relationship between agriculture and the environment, as well as that between certain consumer demands and ecological production.

The degradation of rural life and rural exodus was seen to be essentially the product of a 'serious pathological disorder in our so-called progressive civilization' (European Parliament 1981, p.10). This pathological disorder was seen to arise out of, on the one hand 'the industrial revolution' and processes of urbanisation, which Europe has been subject to over several centuries (European Parliament 1981, p.10). On the other hand, there was the view that the biology of man, which entailed that '[g]iven his height, his build, his mobility, his respiratory system, he needs to have around him a certain space, he must not be rationed as to the amount of unpolluted air he breathes in, he must not be deprived of light, he must be able to exercise his muscles' (European Parliament 1981, p.11). Although, the conception that rural exodus and degradation is caused by a fundamental pathological disorder between the nature of man and processes of industrialisation and urbanisation is not authorised and institutionalised within the CAP, rural exodus is – by the end of the period in question – endorsed as a problem that the CAP needed to address.

Types of solutions proposed in relation to rural exodus involved support to part-time farmers and agri-food industries in rural areas, improving investments in rural areas (by setting up a European Bank for Rural Activities and a Rural Land Development Bank), special initiatives directed at young people in rural areas, and non-agricultural activities such as rural tourism and craft trades (European Parliament 1981, pp.6–8). In addition, a call was made for 'due attention [to] be paid, in the context of the CAP to the legitimate interests of nature, environmental and animal conservation on the one hand, and the supply of healthy, high quality foodstuffs for the population on the other' (European Parliament 1981, p.6). Accordingly, and in order to counteract 'increasing uniformity in agricultural produce', it was suggested that initiatives should be taken to set up legislation to 'guarantee the origin of "ecological sound" products' and European quality labels should be introduced in order to identify certain production methods and the geographical origins of particular agricultural products (European Parliament 1981, p.6). It was argued that 'without wishing to limit the farmer's freedom to choose his own production methods, we may assume that "ecologically sound" products are fairly widespread in the Community and popular with consumers but that their future development could be jeopardized by the unfair commercial practices whereby it is suggested that a product is "ecologically sound", although in fact it is not' (European Parliament 1981, p.20).

As it appears, it is the EP Committee on Regional Policy, which gave voice to 'ecologically sound' agricultural production as a potential and partial solution to some of the problems in agriculture within the auspices of the CAP. The EP Committee on Regional Policy may thus be characterised as exercising policy entrepreneurship in the sense that this Committee contributes to the translation of the conceptions that 'ecologically sound' products are demanded by consumers, and the establishment of a link between agricultural production methods and quality food products within the auspices of the CAP. This policy entrepreneurship is, among other things, made possible by the EP at large, the Council and the Commission. While all three bodies carry and endorse the link as first voiced by the EP Committee on Regional Policy, the particular view of both the Council and the Commission is that the link between organic farming and food quality needs to be confirmed by further research. The EP Committee on Regional Policy also exercised policy entrepreneurship by establishing a forum for communication in the form of an 'own initiative report'.

It is worth noting too, that it was the EP Committee on the Environment, Public Health and Consumer Protection (hereafter the EP Committee on the Environment) which, in an attached letter to the main report, made a call for the link between the CAP, on the one hand, and the protection of the environment, healthy food production and food quality, on the other hand, to be inserted in the EP resolution. This is, in fact, the only concern expressed by the EP Committee on the Environment. The EP Committee on the Environment argued that the report elaborated by the EP Committee on Regional Policy 'has changed its title and its content with the effect that it no longer deals with the ecological and consumer protection implications of the subject, as originally intended'

(European Parliament 1981, p.34; Letter attached by the Chairman of EP Committee on the Environment). Hence, although the EP Committee on the Environment inserted a link between the CAP and the environment within the orbit of the CAP, this Committee also declined an invitation to voice an opinion and contribute to the production of discourse, for instance, in relation to 'ecologically sound' production at this point in time.

Finally, the Commission was explicitly asked to consider the introduction of a common set of rules for organic farming in Europe in the early 1980s. Hence, an early initiative to debate the issue of organic farming in the context of the CAP appeared in the form of a 'Written Question' to the Commission in late 1980 from a Belgian MEP (Ernest Glinne). Referring to a recent agreement to adopt of set of common rules for organic food production among 16 producers, processing and distribution firms in France, the Commission was called on to prepare the harmonisation of organic production throughout the Community. This is, thus, an early identification of organic farming as a separate agricultural sector and a potential subject for political regulation. That is, an MEP's actions may be identified as the exercise of policy entrepreneurship by contributing to the translation of organic farming as a potential sector for community harmonisation and regulation as well as by establishing a forum for communication in the form of a parliamentary question to the Commission (European Parliament 1980; *Agra Europe* 20.3.1981).

In the Commission reply organic farming was linked to ongoing Community financed research into the adverse effects of agro-chemicals on the environment, the reduction of the use of fertiliser and a research programme, with the task of looking into how intensive agricultural production affected the quality of agricultural products. Still, organic farming was at that time considered only of minor importance in the Community and the Commission held that it should not be treated as a separate sector but, rather, left to develop independently of political intervention (European Parliament 1980; *Agra Europe* 20.3.1981). Finally, it is important to note that the Commission – at this point in time – considered research into food quality of relevance for the issue of organic farming: however, the link under investigation is the one between intensive agriculture and food quality rather than one between organic farming and food quality.

Organic Farming as an Issue for Further Research (mid-1980s)

A communication elaborated by the Secretariat-General and for which the Commissioner for Agriculture (Dalsager) was responsible, was adopted by the Commission in 1983. The problems and their causes, as identified in this communication, represent a continuation of those given voice to early on in the ongoing period by the Commission and the EP. Problems of surplus production, budget pressures exerted by agricultural expenditures, and the disfavouring of certain rural regions of the Community, were the central issues prioritised for attention by the CAP (Commission 1983, pp.3–4). Factors of economic recession and high inflation were still being cited as sources of the ongoing problems in agriculture and the introduction of co-responsibility was still a preferred solution (Commission 1983, pp.3–4, 8).

However, some of the solutions formulated just a few years earlier, were now rendered as less viable to resolve problems central to the CAP. For instance, exports to third countries as a potential way to deal with surplus production and the existence of 'effective demand' on the world market was considered uncertain and depending on 'economic growth and credit possibility' (Commission 1983, p.5). Moreover, even though an improved socio-structural policy had already been suggested previously as a potential solution, it was now emphasised that if the Community was to find enduring solutions to the problems of the rural areas 'it must put relative more emphasis on the long term structural action, rather than on market intervention and price support' (Commission 1983, p.40). Finally, it was thought that the CAP should have a broader focus not only on agricultural production but also on 'industries upstream and downstream from agriculture itself' and '[i]n modern economic conditions, a common agricultural policy can hardly exist except within the broader concept of a common food policy' (Commission 1983, p.9).

Technological progress was still conceived in terms of a broader societal development, which the agricultural sector and, by association, the CAP, must adapt to. However, unlike the ideal conception articulated by the Commission in the early 1980s, technological progress was also regarded as a central source of agricultural surplus production. That is, on the one hand '[t]he adaptation necessary in European agriculture is only part of the general adaptation of our society, faced with technological progress and a rate of economic growth lower than in earlier years'. And, the 'well-being' of European agriculture 'can be ensured only by a better integration into the economy as a whole, not by its isolation from the underlying factors which are affecting modern society' (Commission 1983, p.5). On the other hand, surplus production and, hence, rising agricultural expenditures was seen as caused by 'the advance of technical progress' (Commission 1983, p.3, 5). In fact, access to land was considered less important than the development of new technology in increasing agricultural productivity (Commission 1983, p.5). In that sense, the conception of technological progress articulated at this point resembles that appearing in the context of the Mansholt Plan in the late 1960s rather than that articulated in the early 1980s within the CAP. The links between, on the one hand, the CAP and, on the other hand, the environment, healthy food production and quality food, rural exodus and degradation which, with various intonations, had been voiced by the EP and the Commission earlier in the 1980s, was far from immediately turned into regulatory initiatives. However, in late 1983 the Council decided to launch a series of research projects on issues relating to agriculture, which reflected a series of these links (Council 1983). The issues to be addressed by further research had to do with the increasing costs of the use of energy in agricultural production, the uneven economic conditions among regions in the Community, the use of fertilisers and pesticides in agriculture and, additionally, animal health and food quality. It was claimed that the 'concern to improve the quality of all food products is increasing and research is needed to elaborate and clarify some of the important problems involved' (Council 1983; preamble).

The latter consideration led to calls for research into the improvements of food quality. In particular research should look into the relationship between, on the one hand, intensive and extensive food production methods and, on the other hand, food quality. In this regard, research was also considered into the relationship between organic farming and quality food products (Council 1983; annex; structural problems). To be sure, organic farming was by no means central to the research initiatives nor was it conceived of as being a readily available solution to problems within the CAP. Nevertheless, organic farming had become accepted as a potential solution to growing demands for quality food products, although further research was considered needed to establish a clear link between organic production methods and quality food products.

Towards the end of the current period, a Council Regulation endorsed and authorised an ideal conception holding that agriculture and farmers had a role to play in the protection of the environment and the countryside and suggested that the intensification of agricultural production was not conducive to objectives of environmental protection (Council 1985). It appeared, for instance, that 'aid may be granted to farmers who undertake to farm environmentally important areas so as to preserve or improve their environment'. But also that a 'farmer's undertaking must stipulate at least that there will be no further intensification of agricultural production and that the stock density and the level of intensity of agricultural production will be compatible with the specific environmental needs of the area concerned' (Council 1985; Art. 19). That is, intensive agriculture was neither explicated as a problem per se nor as a source of environmental depletions, however, it does appear that the Council endorsed the conception that there is a relationship between the degree of intensiveness of agricultural production and environmental concerns. As mentioned, the 1985 Council Regulation does not impose legal obligations as to the protection of the environment on Member States but, rather, enables national initiatives to be taken with reference to the objective to protect the environment from further intensification of agricultural production. However, even within the Council, which is generally considered to stand surety of the status quo (see Chapter 1), it was then accepted that environmental concerns were relevant in the context of the CAP and that further intensification of agricultural production was in conflict with such concerns.

Lingering doubts were removed from a series of those ideational developments outlined above, and authorised by the Green Paper 'Perspectives for the Common Agricultural Policy', which was prepared by the DG for Agriculture and adopted by the Commission in July 1985. Hence, the main problems of European agriculture were conceived to be surplus production, budget pressures, rural exodus and, additionally, structural diversification – caused by an impending enlargement bringing in the countries Spain and Portugal – and environmental depletion. As to the latter, it was considered:

[a]s a matter of fact, agriculture has a direct and profound impact on the environment of the European Community' and '[i]n the last decades, agriculture – or at least some important parts of it – has undergone a technological revolution which has profoundly changed farming practices' (Commission 1985, pp.49–50). Moreover: [c]hanges in farming practices and the development of modern agricultural techniques have played an important role in the increase in agricultural activity over the last decades. But they have also been identified as a cause – and sometimes even as a major cause – of the extinction of species of flora and fauna and of the destruction of valuable ecosystems such as wetlands, and in some cases have increased risks of the ground and surface water pollution' (Commission 1985, p.50).

A clear link between the modernisation of agriculture and a series of environmental problems was here established. The articulation of a possible link between the modernisation of agricultural production and environmental depletion within the CAP had been abortive in the mid-1970s. Furthermore, in the previous period, although eventually institutionalised there had initially been some doubts about this relationship within the emerging EC environmental policy, which had led to a call for research into the matter. Within the CAP, any lingering doubts had been removed by 1985 and subsequently it was acceptable to refer to modern agriculture as a source of certain environmental problems. It also appears that modern technologies and even a 'technological revolution' was regarded as the source of the modernisation of agriculture and, hence, a series of environmental problems. That is, whereas technological progress within the CAP in the early 1980s was conceived of as a 'fact of life' to which agriculture needed to adapt, by the mid-1980s technological developments in agriculture was also regarded as a source of problems central to the sector. A particular problem identified was concerned with the drainage of wetlands for agricultural production leading to the destruction or reduction of habitats for wildlife.

This development was still being funded by Member States as well as the Community but 'the question is…whether public aids for this activity are any longer justified, particularly since the Community has passed self-sufficiency for many agricultural products' (Commission 1985, p.51). Hence, it was being questioned whether it was legitimate to use public funds to subsidise activities that contribute to environmental depletion. This argument – somehow turned on its head – also generated a solution within the auspices of the CAP. It was envisaged that agriculture had a role to play as the protector of the environment and 'in our industrialised society, this role is perceived to be increasingly important, and if agriculture were willing to accept new disciplines in this context, society should recognise it by providing financial resources'

(Commission 1985, p.vi). That is, an agricultural sector providing a public good was also deserving of public support and the public good demanded or, put another way, the 'choice of society', is a 'Green Europe' (Commission 1985, p.ii).

Finally, in late 1985, after consultation with the EP, the Council, and other interested parties following the Commission Green Paper, the Commission adopted a communication elaborated by the DG Agriculture, which largely reiterated the conceptions articulated in the Green Paper (Kommissionen 1985, pp.3–5, 25–27). Although biotechnology in the long term was expected to contain potential market opportunities for certain agricultural products, technological progress was still conceived of as a source of surplus production (Kommissionen 1985, p.15). Moreover, whereas the Green Paper suggested less intensive and alternative agricultural production methods as potential and partial solutions to problems of environmental depletion and surplus production, the subsequent Commission communication gave voice to biological food production as a potential and partial solution to certain problems within the CAP. Biological food production was articulated as an issue for the CAP to deal with in order to secure the free movement of these goods within the Community as well as to meet a demand from European consumers, which would otherwise be met by third countries. The labelling of biologically produced goods was seen as a preferred solution as it would secure both consumer trust in these products and legitimise higher producer prices (Kommissionen 1985, p.13).

Altogether, the cognitively instituted problems that related to surplus production and budget pressures and, hereunder, the use of public funding to support the already better off farmers identified in the early 1980s, were still in force in the mid-1980s. However, alongside these concerns, a series of more normatively instituted problems had been institutionalised within the CAP by the end of the period. Although all of these concerns were not immediately turned into legal obligations, problems relating to rural exodus, structural diversification, intensive agriculture, and environmental depletion were, by the end of the period, institutionalised as rational concerns within the CAP. Moreover, additional structural and thematic sources of the current problems of the CAP had also been institutionalised by the end of the period. That is, henceforth, it was also rational to cite the modernisation of agriculture, technological progress and a choice of society in favour of a 'Green Europe' as the sources of certain problems within the CAP.

Likewise, additional solutions accepted as legitimate by the end of the period are solutions striving towards a CAP that is more closely integrated into the larger economy as well as an agricultural sector that operates in ways to protect the environment. Finally, by the end of the period, organic farming had become articulated as an acceptable and affordable, yet not readily available, solution and further research was seen to be needed into establishing its potential. Essentially, 1985 not only marked the introduction of a voluntary environmental scheme into the CAP, but it also marked the end of a period running from the early 1980s during which certain ideal conceptions of problems, their causes and solutions that related to, among other things, the protection of the environment, underwent institutionalisation within the auspices of the CAP (see also Table 5.1).

Alternative Conceptions of Agriculture (1978–1985)

It is a commonly held view to see the CAP as a highly sectorised policy field, largely impervious to outside influence (Swinbank 1989; Grant 1993; Nedergaard *et al.* 1993; Lenschow and Zito 1998; Skogstad 1998; Daugbjerg 1999; Pappi and Henning 1999; Ackrill 2000; Sheingate 2000). It has, however, also been proposed that the perceived 'negative externalities' of the CAP, including the ever-increasing expenditures, surplus production and adverse effects of agricultural production on the environment have, to some degree, redefined the expectations of what the CAP should deliver (Nedergaard *et al.* 1993; Patterson 1997; Daugbjerg 1999; Ackrill 2000; Sheingate 2000; Hennis 2001; Greer 2005).

In support of the latter view it is proposed that the ideas translated and institutionalised alongside other concerns within the CAP during the period in question have been made available partly by the EC's environmental policy (see Chapter 4) and partly by proponents of alternative agricultural production methods (see below). Whereas both the EC environmental policy and proponents of alternative agriculture make a link between intensive agricultural production and environmental problems available for translation, it is proponents of alternative agricultural production methods, which supply organic farming as a potential and partial solution to certain problems within the CAP.

It is also a common view that the Commission is the most likely agent to act as a policy entrepreneur and, hence, to facilitate change within the CAP (Patterson 1997; Coleman and Tangermann 1999; Daugbjerg 1999; Sheingate 2000; Fouilleux 2004). Although the EP at large, the Commission and the Council have contributed to the carrying of concepts and conceptions linking organic farming to the CAP during this particular period, it has already been proposed above that the EP Committee on Regional Policy exercised the more vigorous type of policy entrepreneurship by contributing to the translation of organic farming within the CAP. In the following, it will additionally be proposed that agents within alternative agriculture exercised policy entrepreneurship by making ideas available and contributing to the translation of organic farming within the CAP.

Organic Farming Within the CAP (late 1970s)

The first articulations of organic farming as a potential and partial solution to certain problems within the CAP can be traced back to the late 1970s. In the late 1960s, the then Commissioner for Agriculture, Sicco Mansholt, gave his name to the first attempt to carry through a major CAP reform and, in general, the name of Mansholt has been connected with solutions related to the modernisation of agriculture by means of industrialisation and rationalisation of agricultural production methods (*Agra Europe* 19.10.1979; Mansholt interview 1978, p.7)[11].

However, the retired Mansholt had, in the late 1970s, become a member of the Council of The Ecological Agricultural Foundation[12] and began speaking out in favour of keeping the remaining labour force in the countryside and adopting policies aiming to maintain small farms. With reference to the then ongoing energy crisis and economic stagflation, the problems to be addressed were those of high unemployment and the concentration of capital. The free movement of capital, a steep rise in energy prices, and the increasing use of technology rather than labour contributed to these problems. More specifically, in relation to agriculture, the continuous industrialisation of agricultural production involving developments towards larger farm units became a central concern since it forced labour out of agricultural production, which in turn is exerted ever more pressure on public expenditures in the form of unemployment benefits. Two additional problems relating to intensive agricultural production and the use of agri-chemicals were identified. Intensive agricultural production was considered to disturb the ecological balance as well as being unprofitable in the long term. The latter was based on the assumption that a future shortage of raw material – as predicted by *The Limits to Growth* – would drive up the price of artificial fertiliser, making its use uneconomic (Mansholt interview 1978, p.1, 6, 13–14).

The problems identified regarding the agricultural sector were explicitly linked to a depressed economic environment as well as to the 'world problematique' as it was first expressed in the context of the *The Limits to Growth*. For instance, in accordance with the ideal conception of the 'world problematique', little trust was placed in technological progress as a solution to the impending problems in agriculture. It was argued at the time that 'the

---

[11] The interview with the former Commissioner for Agriculture Sicco Mansholt first appeared in a German edition of the IFOAM Bulletin in 1978. In 1979 a full English translation of the interview appeared in a publication by the Soil Association (the national UK association for organic farming) along side a reprint of the 'Conclusions drawn by the European Environmental Bureau from a seminar on the Common Agriculture Policy'. Both the interview and the 'Conclusions' will be referred to individually and by the year when they were first published.

[12] The Ecological Agricultural Foundation is established in 1975 by suggestion of E.F. Schumacher, the author of *Small is Beautiful* (1973) (Mansholt interview 1978, p.1).

methodology of today is based on the view of agriculturalist and biologists who assume a limitless growth of technology. In actual fact we are already doing violence to the nature in many respects and therefore have arrived at the limits of possibilities for increasing production' (Mansholt interview 1978, p.1). That is, the conception of technological progress articulated by the former Commissioner for Agriculture fitted in with the equation of the 'world problematique' as given voice by the 'Club of Rome' in the mid-1970s. Yet, this conception is also in conflicts with the mutation that the 'world problematique' had been subject to in the context of the emerging Community environmental policy, which entailed that the resolving of environmental problems should be pursued through technological innovations. To be sure, the analytical lines drawn between the articulations of certain ideal conceptions are not conditioned by explicated references, yet Mansholt, in fact, refers to *The Limits to Growth* and, indeed, had personal discussions with some of those involved in the 'Club of Rome' (Mansholt interview 1978, p.5, 7).

The solutions to some of the problems within European agriculture as articulated by Mansholt envisaged a CAP that encouraged small farm units, less intensive productions methods and in this regard also organic farming through subsidies. In particular, organic farming was seen as a potential solution to the rising expenditures of intensive agricultural production on energy use and agrichemicals. Moreover, whereas it was doubtful if organic farming would address the problem of agricultural surplus production in Europe, organic farming was conceived to counteract the adverse environmental effects of intensive agriculture and, essentially, contributed to the ecological balance that is needed for sufficient food supplies in the long term (*Agra Europe* 19.10.1979; Mansholt interview 1978, p.4, 13).

The diagnosis of the central problems of the CAP as well as the potential solutions outlined above was echoed in the conclusions from a European Environmental Bureau seminar on the CAP in late 1978 (EEB 1978). In addition, the conclusions explicated that it is the CAP, which contributes to 'trends towards specialisation, concentration and industrialisation in agriculture' (EEB 1978, p.3) and, in general, that the CAP ought to support less intensive farming, smaller farms and rural society. One of the means envisaged to counter the shortcomings or undesirable effects of the CAP was 'more active support for "organic" husbandry comparable to that already available for those practising "conventional" farming; and the initiation of a major research effort to evaluate organic methods and the energy use in different systems of husbandry' (EEB 1978, p.3). Hence, whereas the CAP had been identified as the prime source of the current problems of industrialised agriculture, a changed CAP may also address these problems – among other ways – by supporting organic farming and initiating complementary research. That is, the practice of resolving current problems in agriculture, through the CAP, was not called into question.

The conceptions linking organic farming to European agriculture as articulated by Mansholt and the EEB conclusions appear to have been endorsed and, to some degree, wield authority among people involved in organic farming. In the introduction to the publication, bringing the Mansholt interview and the EEB conclusions together, members of the Soil Association were thus asked 'never lose a suitable opportunity of drawing the attention of as wide a circle of people as possible to its existence' and it was recommended that British MEPs should all receive a copy (Soil Association 1979; Introduction). In particular, the Mansholt interview is considered to be of 'outstanding importance'. It is also worth noting that the introduction is written by Lady Eve Balfour who has been included as one of three people who 'initiated the development of organic farming in the UK and provided a powerful stimulus to the founding of the Soil Association in 1946' (Michelsen *et al.* 2001, p.128).[13] The publication also attracted interest outside of the more narrow circles of those involved in organic farming and was reported at some length in *Agra Europe* (19.10.1979).

This broader attention given to organic farming seems, at least partly, to be generated by the fact that it is a concern raised by a former Commissioner for Agriculture and the person who gave name to the first attempt to radically reform the CAP. This point is illustrated by *Agra Europe* (19.10.1979) running the story under the headline: 'The New Mansholt Plan: Small is beautiful'. Likewise, even though the EEB is not part of the formal decision-making structures of the EU and it is considered to have had only a marginal influence on EU policies, the EEB was set up in 1974 with the support of the Commission with the aim of mediating contacts to organised environmental interests and has thus been consulted regularly on general matters relating to the environment (Grant 2000 *et al.*, pp.52–53; Weale *et al.* 2000, p.106; McCormick 2001, pp.117–118). Against this background, while organic farming was far from being widely accepted and conceived of as constituting a readily available solution to problems within the CAP, it did appear to be accepted among agents involved in alternative agriculture at the margin of the CAP in the late 1970s and was, within this sphere, articulated as an alternative, potential solution to some of the problems within the CAP.

In all, the problems of European agriculture and the CAP, as made available for translation by those involved in alternative agriculture, had to do with processes of industrialisation and concentration in agriculture, the disturbance of an ecological balance, and agricultural surplus production. The causes of the problems made available for translation were multiple, clearly structural and thematic in nature (economic recession, energy crisis, unemployment, technological progress, free movement of capital, intensive farming) and found both inside (intensive farming) and outside (economic recession, energy crisis, unemployment, technological progress, free movement of capital) the CAP. Importantly, the ideal conception holding that technological progress does not supply viable solutions to problems in agriculture in the long term was made

---

[13] See also a special memorial issue of The Journal of the Soil Association, *Living Earth* (1990) Lady Eve Balfour – A tribute to an organic pioneer and the founder of the Soil Association

available for translation by those involved in alternative agriculture, and not through the emerging EC environmental policy (see Chapter 4). The solutions envisaged and made available by those in alternative agriculture had to do with a CAP, which encourages smaller farm units, engages in less intensive farming and, in this context, organic farming.

Organic farming was accepted as a solution since it reduces the use of energy and agri-chemicals, and contributes to an ecological and economic balance needed for a sufficient supply of food in the long term. While organic farming is an acceptable solution to certain problems within the CAP, it is not a readily available solution and research is needed into the scope and potential of the organic farming sector. Finally, those individuals involved in alternative agriculture may be said to have exercised policy entrepreneurship in the sense that they contributed to the translation of organic farming as a potential and partial solution to some of the problems within the CAP. The EEB may also be identified as exercising policy entrepreneurship by establishing a forum for communication in the form of a seminar on the CAP.

The problems articulated within alternative agriculture may be characterised, as normatively instituted in the sense that they are based on a coupling between actual but also predicted further industrialisation and concentration in agriculture and, on the other hand, an ideal conception holding that there exists such a thing as an ecological balance, which public policies should strive towards. Finally and importantly for subsequent discursive and institutional changes within the CAP, the problems related to the modernisation or industrialisation of agriculture, to which organic farming is offered up as a potential and partial solution, is increasingly conceived of as being a central concern within the CAP. That is, even though organic farming was given voice only at the margin of the CAP in the late 1970s, the problems, and some of the sources of the problems to which organic farming is conceived to constitute a potential and partial solution, fit in with some of those articulated closer to the heart of the CAP towards the mid-1980s.

Organic Farming and Policy Entrepreneurship in Member States (1980–1985)

In most Community Member States, organic farming was not attracting attention in this particular period. Yet, in France and towards the end of the period in the Netherlands and the UK, the first sporadic links were established between organic farming and agricultural policy objectives in these countries and occasionally links were formed with the CAP. The first – albeit abortive – request for a Community regulation on organic farming appeared within the EP with reference to the establishment of a set of common rules among a group of organic farmers in France (see above). While the establishment of a set of common rules among a group of organic farmers in France acts as an early example of the appearance of a regulation at the local level of organic production, organic farming was also linked to agricultural policy objectives in the French media at the national level. In Le Monde, French agricultural policy objectives were expressed as being concerned with, on the one hand, the efficiency and international competitiveness of the agricultural sector and, on the other hand, the need to provide employment, a decent living standard for farmers, and to avoid rural exodus. Organic farming was here conceived of as an opportunity for smaller farm units to avoid large capital investment related to intensive agricultural production and as representing a labour intensive production method (*Agra Europe* 8.5.1981). Although a link between organic farming and the CAP was not established, this was an early formulation of organic farming as a partial and potential solution to some of the problems facing French agriculture.

Organic farming also received attention in both the Netherlands and the UK towards the end of this period. In the Netherlands, a group of organic farmers presented a petition to the Dutch Ministry for Agriculture, which called for the EC to exempt organic milk from a super-levy on milk production (*Agra Europe* 29.6.1984). That is, organic farming was also here being formulated as a separate sector and a potential subject to Community regulation. In the UK, the House of Commons Agriculture Committee debated organic farming and its potential in the development of a competitive UK agricultural sector. Unlike in the Netherlands, organic farming was not immediately linked to the CAP but rather seen as a potential source of profitable niche production for UK farmers. However, in the UK, organic farming was linked to Europe insofar as certain other European countries was considered as more advanced in dealing with environmental problems related to agriculture and especially on research into organic farming (House of Commons 1985). Finally, whereas a link between agricultural policy objectives and organic farming in France was articulated in the national media, and in the Netherlands by petitioning the Ministry for Agriculture, organic farming in the UK was under consideration as a solution to a number of agricultural policy objectives in a formal forum within the national political system. While none of the national agents had contributed to the translation of organic farming within the CAP, the Group of Dutch organic farmers may be identified as having exercised policy entrepreneurship in the sense that they contributed to the carrying of organic farming as a potential separate sector and subject for public regulation as well as the linking of organic farming to problems within the CAP. Moreover, the UK House of Commons Agriculture Committee may be identified as exercising policy entrepreneurship by contributing to the carrying of a link between organic farming and certain consumer demands.

Translations, Institutionalisations and Policy Entrepreneurship (1978–1985)

The development in problems, their sources, and solutions as articulated within and around the CAP in the period from 1978 to 1985 may be summed up as shown in Table 5.1. The most significant change in the period from 1978 to 1985 was related to the translation and institutionalisation of environmental problems and solutions within the CAP. Hence, problems linking intensive agricultural production to the depletions of the environment have been translated and institutionalised within the CAP alongside other concerns by the end of the period. In turn, processes of agricultural modernisation and technological progress have been translated and institutionalised as the sources of these problems. It is important to note, however, that while further intensification of agricultural production was endorsed by the Council as a matter of concern in relation to the environmentally sensitive areas, a clear-cut causal relationship between intensive agriculture and environmental depletion had still to be articulated within the Council.

By the end of the period, organic farming had been articulated within the CAP as an agricultural sector, which should be subject to harmonisation and regulation, acting as a solution to certain problems within the CAP. Further, voice was given to a link between organic farming and the production of quality food products. Moreover, whereas organic farming was not an issue within the CAP at the beginning of the period, organic farming had been articulated by the end of the period as an acceptable and affordable solution to certain consumer demands. Yet organic farming is not a readily available solution since further research is needed to establish a clear link between organic farming and food quality. That is, whereas the conception that organic farming is in demand among consumers is the most widely endorsed, it cannot be characterised as institutionalised within the CAP since research is needed into the 'real consumer benefits' of organic food products. In that sense, it might be said that organic farming by the end of this period resembled a linguistic field (cf. Kjær 1996) in which further articulations could take place, and which still had not been authorised and linked to sanctions and, hence, had not at this stage obtained an institutionalised form.

The translation of the problems, their causes and solutions, which have been institutionalised alongside other concerns within the CAP by the end of the period, seem to be conditional on ideas made available partly by EC environmental policy and partly by the alternative conceptions of agricultural problems, their causes and solutions that had been developing in parallel with, but outside of, the CAP.

Table 5.1: Problems and Solutions: The CAP and Organic Farming (1978–1985)

	The CAP (early 1980s)	Alternative Agriculture	The CAP after translation (mid-1980s)
Problems	• Surplus production and budget pressures • The use of public funds to already 'better-off' farmers	• Industrialisation of and concentration in agriculture • Disturbance of ecological balance • Surplus production	• Surplus production and budget pressures • Rural exodus, structural diversification, intensive agriculture, environmental depletion
Causes	• Guaranteed prices and direct product subsidies • Economic recession, energy crisis • Natural and structural condition of agriculture • Forthcoming Community enlargement	• Economic recession, energy crisis, unemployment, technological progress, free movement of capital, intensive farming	• Modernisation of agriculture • Economic recession • Technological progress • Community enlargement • Choice of society in favour of a 'Green Europe'
Type of Problem and Causality	• Cognitively instituted problems • Multi-causal • Structural and thematic causes • Combination of causes inside and outside of the CAP	• Normatively instituted problems • Multi-causal • Structural and thematic causes • Combination of causes inside and outside of the CAP	• Cognitively instituted/normatively instituted problems • Multi-causal • Structural and thematic causes • Combination of causes inside and outside of the CAP with an emphasis on the latter
Solutions	• Co-responsibility • Protectionist CAP • Modernisation of agriculture in certain regions (socio-structural policy) • Technology as a 'fact of life'	• CAP regulation in favour of smaller farm units and less intensive farming • Technological progress not viable in the long term	• Agriculture as protector of environment • CAP dealing with agriculture as part of larger economy (Common Food Policy)
Organic Farming as a Solution	• Should be left to develop independent of political intervention	• Reduce input of energy and agri-chemicals, contributes to ecological and economic balance needed to ensure sufficient supplies of food in the long term (acceptability) • Doubtful if addresses surplus production problem • Research needed on current scope and potential of organic farming sector (not readily available)	• Meeting demand from consumers (acceptability) • Labelling needed to ensure trust and legitimise premium prices (affordability) • Research needed on link between organic farming and food quality (not readily available)

The translation of problems leading to the establishment of a link between intensive agriculture and environmental depletion, and the conception that organic farming is a potential and partial solution to such problems seems, hence, both to be conditional on ideas made available by the EC environmental policy and by ideas made available by those involved in alternative agriculture. However, whereas the identification of processes of modernisation of agriculture, as the source of environmental problems, appeared to draw on ideas made available by the EC environmental policy, the identification of technological progress as a source of rather than a solution to current problems, appeared to draw on ideas made available by those involved in alternative agriculture.

Essentially, whereas the CAP undoubtedly is a highly sectorised policy, it would appear that certain ideal conceptions made available by the emerging EC environmental policy, and by those involved in alternative agriculture, had been selected for translation and institutionalised within the CAP during the period from 1980 to 1985. This development seemed to be enabled not only by the widespread acknowledgement of the presence of an energy crisis and economic recession and a perceived need for change, but also by some degree of fit between the ideal conceptions articulated within and around the CAP, particularly on what were seen as problematic issues within the CAP. To be sure, the exercise of policy entrepreneurship vis-à-vis organic farming was conditioned by other ideational developments such as, for instance, the institutionalisation of the conception that intensive agricultural production has effects on the environment and that the CAP should strive towards counteracting such effects. That said, the exercise of entrepreneurship vis-à-vis the articulation of links between organic farming and the CAP in the period from 1978 to 1985 may be summed up as in Table 5.2. Essentially, the more vigorous type of entrepreneurship, which contributed to processes of translation, were exercised by the EP Committee on Regional Policy, as well as by those involved in alternative agriculture. It is, however, also worth noting that the EP Committee on the Environment and the Commission did not contribute the establishment of linkages between organic farming and the CAP during the period under investigation.

Table 5.2: Policy Entrepreneurship and the Formation of a Linguistic Field concerned with Organic Farming (1978–1985)

Concepts and conceptions	Types of policy entrepreneurship			
	Translators	Establishing a forum for communication	Carriers	Non-carriers
Organic farming as a sector for political regulation	• Individual MEP	• Individual MEP (Parliamentary question)	• Group of Dutch organic farmers	• Commiss ion
Linkage of organic farming to problems within the CAP	• Agents involved in alternative agriculture • Individual MEP; EP Com. Reg.	• EEB (Seminar on the CAP) • Individual MEP (Parliamentary question); EP Com. Reg. (Own initiative report)	• EP • Group of Dutch organic farmers	• EP Com. Env.
Linkage of organic farming to food quality	• EP Com. Reg.	• EP Com. Reg. (Own initiative report)	• EP • Commission; DG Agri. • Agriculture Council (yet further research needed to confirm link)	• EP Com. Env.
Linkage of organic farming to consumer demands	• EP Com. Reg.	• EP Com. Reg. (Own initiative report)	• EP • Commission; DG Agri. • Agriculture Council; UK Agri. Com.	EP Com. Env.

6

Conflicts Over Meaning and Policy Entrepreneurship within the CAP (1986–1992)

The MacSharry or 1992 reform is broadly agreed to have marked the greatest departure from the status quo in the course of the development of the CAP (see Chapter 1). A common explanation offered for the adoption of the 1992 reform is that it occurred in the context of external pressure from, most notably, the parallel negotiations on trade liberalisation taking place within the GATT (Patterson 1997; Coleman and Tangermann 1999; Sheingate 2000). Likewise, it has been proposed that concerns about the Community enlargement southward and potentially eastward also contributed to the adoption of the 1992 reform (Nedergaard *et al.* 1993; Lenschow and Zito 1998; Ackrill 2000). While not discounting the existence of concerns during this particular period about both obligations related to the GATT negotiations and the enlargement to the south and other developments commonly identified as external to the CAP, it is proposed here that problems central to the CAP are increasingly explained by sources identified as endogenous to this field compared with the period of time up to 1985. For instance, tensions with trading partners, a decline in the value of the US dollar, and falls in world prices on agricultural produce, is part of the discursive and institutional context of the CAP during the period from 1986 to 1992. Yet, most notably, there is a renewed prominence given to the favoured regulatory mechanisms of the CAP. That is, resembling the early 1980s, an emphasis is put on guaranteed prices for agricultural produce and direct product subsidies and, unlike the early 1980s, intensive farming is identified as the prime source of problems central to the CAP. Additionally, the CAP is commonly considered to be governed largely by the Council of Agricultural Ministers, the Special Committee for Agriculture, the DG for Agriculture, and the farm lobby, which together are thought to safeguard the continuity of the CAP (Swinbank 1989; Grant 1993; Nedergaard *et al.* 1993; Lenschow and Zito 1998; Skogstad 1998; Daugbjerg 1999; Pappi and Henning 1999; Ackrill 2000; Sheingate 2000). At the same time, the increased formal decision-making powers attributed to the EP by the adoption of the SEA in 1986 and the TEU in 1992 has been suggested to be potentially favourable to change within the CAP (Nedergaard *et al.* 1993). Although empirical research on the matter is otherwise scarce, it has most recently been proposed that the EP at large, individual MEPs, various EP Committees and in particular the EP Committee on the Environment increasingly has been giving momentum to certain changes within the CAP (Roederer-Rynning 2003). While not necessarily ascribing the contributions of the EP to its increased formal decision-making powers, it is proposed here that the EP and groupings within the EP have given momentum to processes of institutionalisation in the case pursued, that is, the articulation and institutionalisation of organic farming within the auspices of the CAP.

Moreover, contrary to the idea of this being a phenomenon principally located in the later part of the 1990s (Roederer-Rynning 2003), it is suggested that the EP gave momentum to institutional change prior to this. That is, whereas the EP Committee on Regional Policy in particular contributed to the emergence of what may be characterised as a linguistic field, which links organic farming to the CAP in the early 1980s, various groupings within the EP also contributed to the translation and institutionalisation of organic farming within the CAP and, hence, exercised policy entrepreneurship during the period from 1986–1992. In general, it is suggested that the period from 1986 to 1992 may be characterised largely by a continuation of the cognitively and normatively instituted problems institutionalised by the end of the previous period, and by a specification of the solutions preferred, again by the end of the previous period.

This particular period may, however, also be characterised by, initially, conflicts over the degree of priority to be attached to both cognitively and normatively instituted problems and by some ambiguity as to the sources of those otherwise familiar problems, and an eventual mutation of these problems and a specification of their sources. Importantly, it is proposed that intensive agriculture is institutionalised as a source of both cognitively and normatively instituted problems and, hence, a mutation of such problems is evident within the CAP by the end of the period. It is also suggested that it is during this period that organic farming became institutionalised within the auspices of the CAP as an agricultural sector for Community regulation and as a solution to certain problems within the CAP.

Further, during the investigated period, it is suggested that processes of articulation and institutionalisation and, hence, institutional change is given momentum by a series of conflicts over meaning and the exercise of policy entrepreneurship. Conflicts over meaning, for instance, are found within the EP, the Commission and the Commission Services as well as between the EP and the Commission. With regard to the institutionalisation of organic farming, as carriers the EP, the Commission and the Commission Services and a number of Member States all contribute to such processes during the current period. It is, however, proposed that the EP Committee on Agriculture, Fisheries and Food (hereafter the EP Committee on Agriculture) and, particularly, individual MEPs and the EP Committee on the Environment have exercised the more vigorous type of policy entrepreneurship by contributing to the translation of organic farming within the CAP. It is these discursive and institutional developments, which will be considered below by consulting a total number of 55 empirical documents and articles.

Conflicts Over Meaning Within the EP (mid-1980s)

Enabled by the translation of an ideal conception, which linked intensive agricultural production to certain environmental problems, a number of environmental problems and solutions had been institutionalised alongside other concerns within the CAP by the end of the previous period. That is, after 1985, problems, their sources and solutions within the CAP not only came to evolve around issues of, for instance, surplus production, budget pressure and the distribution of public support among farmers and regions, but also around intensive agriculture and environmental depletion. To be sure, although intensive agriculture became institutionalised as a concern within the CAP, most notably the Agriculture Council has yet to endorse a causal relationship between intensive agricultural production and environmental depletion. Still, the institutional developments of the previous period, essentially, introduces an additional number of lines of conflict over meaning as to, for instance, the priority to be attached to, and the scope of, environmental problems, the particularities of their sources, and solutions in the context of other problems, sources and solutions.

   Such conflicts over meaning are, for instance, found in the EP response to the 1985 Commission Green Paper. The EP responded through an 'own initiative' report and a motion for a resolution, which was drawn up by a group of Green MEPs within the EP Committee on the Environment (Jones and Clark 2001, p.69), and an attached opinion was issued by the EP Committee on Agriculture. The motion for a resolution adopted against the background of the 'own initiative' report was voted for by 23 of the members in the EP Committee on the Environment with two other members choosing to abstain: it was subsequently endorsed by the EP in early 1986 (Europa Parlamentet 1986). On the one hand, the main report and the attached opinion represent two unlike conceptions as to the sources of the problems facing the CAP and the nature of potential solutions. On the other hand, there also appears to be some degree of fit between the main report and the attached opinion as to the problems facing the CAP even if conflicts existed over the scope of, and priority accorded to, various problems.

Along these lines, the main report identified a series of general problems in agriculture related to pollution, surplus production, the decline of job opportunities in agriculture and, by association, rural depopulation and urbanisation, the decline of land and soil quality, and a lack of food quality affecting human health (European Parliament 1986, pp.13–18). Although it was pointed out that 'agriculture is not the enemy of the environment' and 'intensive farming is not the only cause of the deterioration of the rural environment' (European Parliament 1986, p.13), the current principles of operation of the CAP and its instigation of intensive agricultural production were identified as the cause of environmental problems in rural areas.

The CAP principles embracing 'market unity', 'Community preference' and 'financial solidarity' are thus all at odds with the protection of the environment (European Parliament 1986, p.19). More specifically, the striving for market unity through the unhindered movement of goods and uniform prices across the Community contains a potential 'conflict with environmental interests, because the natural conditions in agricultural production of Europe vary widely' (European Parliament 1986, p.19). That is, one source of environmental problems is being seen as related to the objective of market unity, which is given priority above environmental concerns within the CAP. Another source is found in the preference given to agricultural products produced within the Community, which disregards any possible environmental effects related to the location of various types of agricultural production. Finally, financial solidarity, which implies that the burden of CAP expenditures is shared among Member States, is conceived of as to favour short-term interests above concerns of the environment in the long term (European Parliament 1986, p.19).

It has been suggested that the CAP should, for instance, develop a policy on the use of land, a policy concerned with food quality, a price policy embracing socio-economic and environmental concerns, and a policy on forestry (European Parliament 1986, pp.6–10). Solutions were considered, in general, to be found in a CAP in support of extensive agriculture and it was emphasised that '[e]nvironmental issues must be treated as a *key element bound up with the problems of agricultural surpluses and questions of prices and economic incentives in relation to types of production and production methods*' (European Parliament 1986, p.21; original emphasis). It is along these lines 'biological farming' came to be proposed as a potential and partial solution to some of the problems of the CAP. This type of production method was considered to: counteract the adverse environmental effect of intensive farming, lessen the use of energy in agriculture, improve the quality of land and food products, and also have a positive effect on animal health. Additionally, it was considered that this production method needed a higher input of labour and has a lower output of food products – 'but in the view of the current surpluses in Europe this is hardly a point worth worrying about' (European Parliament 1986, p.36).

Compared with the CAP at the end of the previous period, the report thus identified further problems to be added to those already existing, to which organic farming could be seen as a potential and partial solution. In addition, although the sources of the otherwise familiar and normatively instituted problems were still thematic and structural in nature, a simple mono-causality was established between the principles of operation of the CAP and environmental depletion. The link between the CAP and environmental problems, the priority to be given to environmental problems, and the availability of solutions within the CAP were, however, contested.

In an opinion attached to the main report by the EP Committee on Agriculture it was thus emphasised that the continuing production surplus was the most important problem facing the CAP and – if not addressed – 'the CAP risk[ed] disintegration, which would call into question the whole shape and future of the European Community' (European Parliament 1986; Attached Opinion, p.57). Although, a reform of the CAP may also be seen as addressing environmental concerns, the objective of such a reform had been, primarily, to deal with the increasing agricultural production surpluses and the resultant financial drain on the Community budget (European Parliament 1986; Attached Opinion, p.57). Second, the link between modern agricultural production and adverse environmental effects had been questioned and – even if the link could be shown to exist – the scope of environmental problems related to agriculture remained unclear (European Parliament 1986; Attached Opinion, p.57). With reference to a recent public hearing convened by the EP Committee on the Environment on the issue it was stated that:

[e]xperts talked of damage to the environment as a result of pollution, but there was no clear statement which qualified how much damage was being done by modern agricultural methods. In more general, public debate, there has also been considerable talk of 'agriculture damaging the environment', but again it is difficult to say to what extent this is happening (European Parliament 1986; Attached Opinion, p.57).

Hence, whereas the attached opinion issued by the EP Committee on Agriculture endorsed the cognitively instituted problem of surplus production, it also contained reservations about the normatively instituted problems based on the ideal conception that the CAP should contribute to the protection of the environment and questioned the scientific evidence in support of a link between modern agricultural production and environmental depletion.

Even though the link between modern agricultural production and environmental depletion and the scope of problems in this regard was called into question, the list of problems within the CAP seen as requiring attention otherwise corresponded with those given in the main report (European Parliament 1986; Attached Opinion, pp.61–63, p.67). However, conflict existed on whether or not solutions to these otherwise familiar and largely environmentally related problems were available in a Community context since 'the Community's environmental policy is inevitably limited and that action to reduce the negative effects of agricultural development is relatively small and piecemeal' (European Parliament 1986; Attached Opinion, p.59). It was also questioned as to whether the environmental policy of the Community or, more specifically, the objective of integrating environmental concerns into other Community policies, had been successful (European Parliament 1986; Attached Opinion, p.60). Furthermore, the existence of a link between a restrictive price policy and less intensive farming was not recognised. Rather, the opposite conception was voiced: '[i]t is quite possible that restrictive prices will encourage further intensification of production, and increased environmental damage, as farmers try to compensate for lower real incomes' (European Parliament 1986; Attached Opinion, p.63).

Finally the opinion issued by the EP Committee on Agriculture explicitly refrained from establishing a clear causal relationship between, on the one hand, either the functioning of the CAP or technological development and, on the other hand, environmental depletion. Instead it was suggested that the two causes were probably 'inextricably intertwined' and '[i]f you establish a system which rewards high outputs and high technology, it is really academic whether you blame the system on the instruments of the system' (European Parliament 1986; Attached Opinion, p.61). It is against this 'academic' reservation that environmental problems were identified as appearing 'under the CAP regime' rather than being caused by the CAP per se.

With these reservations in mind as to the existence of a causal relationship between modern agriculture and environmental depletion, the scope of environmental depletion, and the availability of Community solutions, it is not contested that the extensification of agriculture should be encouraged by the CAP. Moreover, it was considered that alternative farming practices should be encouraged by 'facilitating exchanges of views and information on methods of production' rather through financial support (European Parliament 1986; Attached Opinion, p.66). Finally, from the attached opinion, it appears that 'while "alternative" farming deserves to be encouraged in the interests of consumer choice, effects on the employment and the land, it should be allowed to establish its own place in the market, which is growing rapidly, despite the premium price its products command' (European Parliament 1986; Attached Opinion, p.69). That is, while it was considered that alternative farming – not specifically organic farming – should be subject to Community regulation due to its contribution to the maintenance of the countryside, it was also considered that alternative agricultural production methods was not to receive extraordinary financial support.

Policy Entrepreneurship Within the EP (1986–1992)

Regarding the exercise of policy entrepreneurship, which has given momentum to the institutionalisation of organic farming, the EP Committee on the Environment have contributed to the translation of some of the conceptions institutionalised during this particular period, unlike the period immediately prior to this one where this Committee abstained from contributing to the production of discourse on such links. During the period from 1986 to 1992 the EP Committee on the Environment thus contributed to the translation and institutionalisation of organic farming as a potential sector for Community regulation, as a potential and partial solution to certain problems within the CAP, and to linking organic farming to the fulfilment of CAP objectives related to environmental protection and the maintenance of the countryside. Moreover, the EP Committee on the Environment also exercised policy entrepreneurship by establishing forums for communication in the form of an 'own initiative' report as well as by initiating a public hearing (European Parliament 1986).

It is also worth noting that the EP Committee on Agriculture, at the beginning of this period, explicitly questioned the link between modern agricultural production and environmental depletion, the scope of environmental depletion related to modern agricultural production techniques, and also whether solutions to the claimed environmental problems were available within the Community. Yet, this committee contributed to the translation and institutionalisation of links between 'alternative farming' practices – i.e. not specifically organic farming – and issues concerning demands on the markets as well as contributed to the conception that alternative farming could bring about desired effects on the countryside and, in this particular regard, improve employment prospects in designated areas (European Parliament 1986; Attached Opinion).

In addition, by the end of the period, a report on a proposed EU regulation on the organic farming sector drawn up by the EP Committee on Agriculture endorsed a causal relationship between intensive agriculture and environmental depletion. For instance, the report held that '[t]he increasing use of chemicals in agriculture, in particular pesticides, and also the production of such substances involves enormous risks for the environment and for the health of the population in general' (European Parliament 1990b, p.36). It also appeared, however, that '[t]he long-term pursuit of the "slimming down" of European agriculture has resulted in the industrialization of agricultural production which has had adverse effects on the environment, humans and animals' (European Parliament 1990b, p.36). Essentially, the report upheld that restrictive price policies and 'structural changes' had resulted in forced intensive production methods on agriculture and, in turn, led to environmental depletion (European Parliament 1990b, pp.35–36). Nevertheless, towards the end of the period the EP Committee on Agriculture exercised policy entrepreneurship by carrying concepts and conceptions that linked organic farming to the CAP. Thus, organic farming was endorsed as an agricultural sector for Community regulation and as a potential and partial solution to certain problems within the CAP, as well as contributing to the fulfilment of CAP objectives that related to environmental protection and the maintenance of the countryside (European Parliament 1990b, pp.35–37).

However, it should be noted that, above all, the report drawn up by the EP Committee on Agriculture linked organic farming to the support and protection of small-scale farming, a conception which, during this particular period, was not institutionalised (European Parliament 1990b, pp.36–37). Moreover, in a series of written and oral questions to the Commission, MEPs continuously referred to organic farming as a separate agricultural sector that should be subject to Community regulation, gave voice to links between, on the one hand, organic farming and, on the other hand, certain concerns within the CAP – these including consumer demands and the protection of the environment and the countryside. A number of individual MEPs may, therefore, also be identified as contributing to the translation and institutionalisation of organic farming throughout the period from 1986 to 1992. Individual MEPs also exercised policy entrepreneurship by continuously establishing forums for communication through fielding parliamentary questions to the Commission (*Agra Europe* 13.5.1988, 9.6.1989a, 18.8.1989; European Parliament 1987, 1989, 1989a, 1990, 1990a, 1991).

Finally, rather than contributing to processes of translation or the establishment of a forum for communication, the EP at large may, at that time, be identified as having exercised the less vigorous type of policy entrepreneurship, which involves carrying concepts and conceptions that, over time, are institutionalised. Hence, the EP endorsed a motion for a resolution based on a report draw up by the EP Committee on the Environment. Moreover, towards the end of the period, during an EP debate on issues related to organic farming it became – albeit to varying degrees – widely recognised that industrial and intensive agriculture was the source of problems within the CAP, and had to do with the decline of employment opportunities in agriculture and environmental depletion (European Parliament 1991a). Again, to a varying degree, organic farming was conceived as contributing to the production of food products of a high quality, the improvement of consumer health, the fulfilment of consumer demands, the protection of rural landscapes, constituting a market opportunity, and complying with the requirements imposed by the negotiations within the GATT (European Parliament 1991a). Although the links between, on the one hand, organic farming and, on the other hand, food quality and consumer health was not institutionalised during this particular period, the EP contributed to the carrying and institutionalisation of organic farming as an agricultural sector in need of Community regulation, and one that could act as a solution to certain problems within the CAP, as well as meeting consumer demands, and fulfilling objectives related to environmental protection and maintenance of the countryside.

Conflicts Over Meaning Within the Commission (late 1980s)

The institutionalisation of normatively instituted environmental problems by the end of the previous period, in essence, leading to the introduction of an additional number of lines of conflicts – for example, over the level of attention to be attached to environmental problems, the scope of such problems, the exact nature of the sources of such problems, and their potential solutions – may also be illustrated by the ideal conceptions articulated within the Commission in the late 1980s. The conflicts over meaning within the Commission are, however, not as clear-cut following functional divisions as within the EP but, rather, appear as a number of uncertainties in publications forwarded by the Commission.

For instance, the central problem of the CAP – as identified in a communication 'on the prices for agriculture products and on related measures (1987/88)' elaborated by the DG for Agriculture and adopted by the Commission in 1987 – was a persistent and increasing growth in agricultural production. Not only was increasing agricultural output exerting pressure on the CAP budget due to the expenditures used to guarantee agricultural prices against specified areas of production, but also these CAP expenditures represented an increasing burden on the wider Community budget in terms of the SEA implementation costs (Commission 1987, pp.1–2). The accumulation of intervention stocks was considered a particularly serious problem at that time: '[T]he level of these stocks is a constantly unsettling factor which precludes any prospects of lasting improvement on the markets and considerably reduces the Community's bargaining power vis-à-vis importing countries' (Commission 1987, p.2). The problem of surplus production was worsened by similar problems of 'over-production at the world level, the relative inelasticity of supply, and the shortage of foreign exchange in the Third World' (Commission 1987, p.12).

While some of the causes of agricultural surplus production and related problems are internal to the CAP, the causes were generally regarded as lying outside of the CAP and even outside of the Community. Hence, internal markets were seen as 'sluggish' while international markets were seen to be depressed due to the decline in the value of the US dollar, and the adoption of the US 'Food Security Act', which was considered to have resulted in lower prices for agricultural products and which led to a significant downturn in demand from oil-producing countries (Commission 1987, p.8). Furthermore, the preferred solutions were already well known such as a restrictive price policy, more flexible interventions and a stronger emphasis on the co-responsibility of farmers: there was also an acknowledged, although formally unspecified, 'policy for quality' (Commission 1987, p.15).

In other words, it was the familiar cognitively instituted problems that were identified and familiar solutions that were also largely preferred. Although defined differently when compared with the situation by the end of the period prior to this one, the sources of (then) current problems were still seen as multiple, structural and thematic in nature. Likewise, some of the sources of the problems central to the CAP were still regarded as internally located (guaranteed prices, sluggish internal markets) but mainly regarded as externally located to the CAP (over-production at the world level, depressed international markets, decline in value of US dollar, the US Food Security Act).

Against this backdrop, the Commission adopted a Green Paper, 'Environment and agriculture', in June 1988. The DG for the Environment is primarily responsible for the Green Paper but both the DG for Agriculture and the DG for the Internal Market had been involved in its elaboration (Commission 1988). From the Green Paper it appears that agricultural surplus production and environmental depletion were identified as the two central problems facing European agriculture. The normatively instituted problem of environmental depletion had to do with the: deterioration and extinction of flora and fauna, deterioration and erosion of soil, deterioration of water and air, changing landscapes and the decreasing quality of agricultural products (Commission 1988, pp.6–7). Such problems were considered to stem from increasingly intensive production methods within agriculture, which was a development characterised by higher yields as well as 'mechanisation, the use of agrochemicals, [and] the improvement of cultivation techniques' (Commission 1988, p.6).

The issue of higher yields is the core concern of the second cognitively instituted problem of the CAP – agricultural surplus production. Although the development of intensive agricultural production methods is the central cause of both problems of environmental depletion and surplus production, developments outside agriculture such as tourism, urbanisation and industrial pollution have also contributed to the formation of the problems central to the CAP (Commission 1988, p.6). In fact, although intensive farming led to problems of surplus production and environmental depletion it is only an intermediate and symptomatic factor, as intensive farming itself is seen as being caused by 'a technological revolution' in agriculture stemming from 'pressures coming from other economic activities such as urbanisation, industrial development and transport' (Commission 1988, p.1).

In order to address the problems identified, a series of measures in support of the extensification of agricultural production various policies have been put forward concerning the use of land, the use of pesticides, curbing intensive animal and plant production, improving food quality as well as a number of accompanying measures in support of agricultural practices compatible with certain environmental requirements (Commission 1988, pp.9–15). In relation to the latter, research was called for into organic farming with the aim of establishing a set of codes for 'good agricultural practices'. Moreover, organic farming was linked to the production of quality products, considered to have favourable effects on the environment as well as offering a way to diversify the sources of revenue for farmers (Commission 1988, p.15). Thus, organic farming was not conceived of as a mere production method that supplied consumers with quality food products but also as having properties that offered a potential and partial solution to both some of the existing environmental and economic problems in agriculture.

As it appeared, the cognitively instituted problem of agricultural surplus production identified in the Commission communication from 1987 fitted in with the Green Paper from 1988. At the same time, conflicts existed regarding the identified cause of agricultural surplus production. That is, whereas surplus production in 1987 was blamed on guaranteed prices, sluggish internal markets, over-production at the world level, depressed international markets, a decline in the value of the US dollar, and the adoption of the US Food Security Act, the 1988 Green Paper pointed to intensive agriculture and technological progress in agriculture imposed by developments outside the agricultural sector as the causes of surplus production. In addition, the 1988 Green Paper also blamed intensive agriculture and technological progress for a number of environmental problems. In other words, the sources of current problems as identified in the 1988 Green Paper were seen to be structural and thematic but also – and contrary to the multi-causal sources identified in the 1987 Commission communication – simple in the sense that if intensive agriculture could be curbed it would have positive effects on concerns relating to surplus production as well as on the environment. As it turned out, it was in fact the problem causality articulated in the Green Paper, which was institutionalised by the end of this particular period.

From mid-1988 organic farming within the Commission was linked to broader objectives of environmental protection and the improvement of the rural environment. Furthermore, organic farming was established as a niche sector that could bring about a better balance between the supply and demand of agricultural products and, potentially, be a profitable opportunity for farmers (Commission 1988; *Agra Europe* 17.11.1989, 1.12.1989). Such conceptions were largely echoed in a Commission proposal in late 1989 that called for a Community regulation on the organic sector. This proposal was primarily elaborated by the DG for Agriculture with the involvement of, in particular, the DG for the Internal market but also including the participation of DGs for External Economic Relations, the Environment, and Enterprise Policy, Distributive Trades, Tourism, and Cooperatives (Commission 1989, p.1).

The status of organic produce vis-à-vis quality food products, which is a link that was first articulated during the previous period (the late 1980s) was, however, still unresolved within the Commission and the Commission Services at that time. It was thus explicated in 1989 that organic food products should not be considered as being of a superior quality (Commission 1989, p.3) and the outgoing Commissioner for Agriculture (Frans Andriessen) stated that the Commission had no official opinion on whether there were any differences in the quality of conventional and organic products saying: 'it is for the consumer to assess such differences' (*Agra Europe* 9.6.1989). That is, organically produced food products were seen, on the one hand, to be linked to a food quality concern in the Green Paper from 1988, elaborated primarily by the DG for the Environment (cf. Commission 1988). Yet, on the other hand, it had been emphasised in the proposal for a Community regulation, primarily prepared by the DG for Agriculture (cf. Commission 1989) – as well as by the outgoing Commissioner for Agriculture – that organic food products were not in fact superior to conventionally produced food products. Yet, rather than being a conflict that followed functional lines, that is, appearing between the DG for the Environment and the DG for Agriculture, the conflict over the status of organic food products vis-à-vis food quality defied such lines. Hence, it appeared in a communication adopted by the Commission in late 1988 (for which the DG for Agriculture was primarily responsible) that the Commission had a concern with, and wished to further the production of, quality food products: in this context it considered whether a quality label should be introduced for organically farmed food products (Kommissionen 1988, p.2). By the end of this period it was the conception that organic food products are not of superior quality that was institutionalised within the CAP, but the conflicting conceptions reappeared in the subsequent period ranging from 1993 to 2005.

Policy Entrepreneurship Within the Commission (late 1980s)

At the beginning of the previous period, the Commission conceived organic farming as an issue, which should be left to develop independently of political intervention, and thus a concern outside of the CAP. Yet towards the end of the previous period the Commission had put out a call for research into the possible benefits of organic farming. It was against this background the Commission fed expectations of a forthcoming Community regulation on organic production throughout the period from 1986 to 1992. Such expectations were first articulated by the Commission in the context of the CAP in 1986 and reiterated throughout the remainder of the period (e.g. *Agra Europe* 13.5.1988, 9.6.1989, 18.8.1989; European Parliament 1987, 1989, 1989a, 1990, 1990a, 1991).[14]

The expectations of the Commission to forward a proposal for a policy addressing the organic farming sector was also fed by a Commission chief research programme coordinator (Mark Cantley) at a conference on 'green alternatives' to the CAP, which had been convened by the Rainbow Group of Green MEPs in 1988 (*Agra Europe* 11.11.1988). Hence, rather than leaving the organic sector to develop as originally set out by the Commission in the early 1980s, conceptions could be seen to shift, with the Commission clearly indicating a move towards carrying a conception of organic farming as an

---

[14] The European Parliament references include the answer given by the Commission to parliamentary questions.

agricultural sector, which should be subject to Community regulation. Moreover, the Commission endorsed links between, on the one hand, organic farming and, on the other hand, certain concerns within the CAP related to consumer demands, and the fulfilment of CAP objectives related to the protection of the environment and the maintenance of the countryside. Likewise, first the DG for the Environment (in association with other DGs) and towards the end of the period the DG for Agriculture (in association with other DGs and in particular the DG for the Internal market) both contributed to the carrying of these conceptions.

However, the policy entrepreneurship exercised by the Commission, the DG for the Environment and the DG for Agriculture, did not contribute to the connection of concepts and conceptions between fields but, rather, contributed to the endorsement of conceptions given voice previously among others within the EP and, in particular, by the EP Committee on the Environment as well as a number of individual MEPs. In addition, both the response of the Commission to questions raised in the EP and the conference convened by the Rainbow Group of Green MEPs illustrate how the Commission became increasingly involved in the production of discourse on the links between organic farming and the CAP, yet forums for communication was essentially established by various individuals or groups of MEPs.

Organic Farming and Policy Entrepreneurship in Member States (1986–1992)

Parallel with the developments addressed above, organic farming throughout this period received increasing attention in a number of Community Member States, particularly in the UK and Denmark but later on also in the Netherlands, Spain, Germany and France. There are, however, some variations among the Member States as to which particular problems were seen as in part solvable by organic farming.

In the UK, a study commissioned by the 'Cooperative Development Division of Food from Britain' showed that UK sales of organic food was set to grow considerably in the years to come and consumers were willing to pay extra for such food products. It also appeared, however, that the quality of organic produce and production had to be improved. In addition, rather than having two contemporaneous but differing sets of standards attached to organic production methods (one being UK and the other international), it was suggested that one set of international standards for organic production was preferable in the endeavour to improve the distinctiveness of organic produce and, hence, simplify communication to consumers about these products (*Agra Europe* 30.5.1986).

Organic farming was also linked to UK agricultural policy objectives during the current period. For instance, in late 1987, a call appeared from within the UK governing Conservative party for a fundamental change of UK agricultural policy. This call drew on some of the conceptions forwarded during the time when the issue of organic farming was raised in the House of Commons Agriculture Committee two years earlier. Accordingly, the central problems of UK agriculture were identified as rising agricultural productivity and the decline of farm gate prices and, in turn, a further need for farmers to increase productivity and their use of agrochemicals in order to maintain profitability. The Ministry of Agriculture, Fisheries and Food (MAFF) and the CAP were identified as the sources of current problems affecting UK agriculture and the solution suggested was a development of organic or 'green' farming. It was suggested that the farmer should be encouraged 'to farm biologically, to support the wildlife and other aspects, while sustaining agriculture on a sound footing, using both traditional techniques and the benefits of modern research' (*Agra Europe* 6.11.1987). It was also around this time that the Conservative Government set up a UK Register of Organic Food Standards (UKROFS) in order to establish a unified set of organic food standards and, hence, increase consumer confidence (*Agra Europe* 5.5.1989).

In the early 1990s, the BSE scare further fed consumer and supermarket demands for, in particular, organic beef, which was seen as a potentially profitable niche market for British farmers. Consumer interest in animal welfare and the taste of organic products were considered important sources of the rising demand for organic produce. Moreover, the British government, MAFF and the National Farmers Union (NFU) demonstrated an increasing interest in the development of a market for organic food products. However, the current supply of organic food products was unable to meet the rise in demand, and the standards for governing organic farming were still in dispute (*Agra Europe* 26.1.1990, 24.8.1990, 12.10.1990, 1.11.1991). Altogether, the CAP, MAFF and intensive farming were identified as the sources of the problems in UK agriculture. Yet, organic farming as a solution was not immediately linked to the CAP and, in fact, the UK government responded 'coolly' to the introduction of a Community regulation on organic farming since the Soil Association and UKROFS had already introduced their own rules and it was thus claimed that a Community regulation would be less stringent (*Agra Europe* 18.8.1989).

As in the UK, a growing interest in organic food products and a preparedness to pay a premium price for these products was noted among consumers in Denmark at the beginning of this particular period. A recent public debate concerned with the effect of intensive farming on, in particular, water pollution – as well as a wish to develop alternative home and exports markets – have together contributed to the increased level of attention given to organic farming. This rising interest in organic food products had also been reflected in the employment of the first consultants on issues related to organic farming within the two largest farmers' unions – the Danish Smallholders' Union and the

Danish Farmers' Union (*Agra Europe* 19.9.1986). In mid-1987, the first national regulation of organic production in Europe – including a national support scheme for conversions to organic farming – was adopted by a large majority in the Danish parliament (*Agra Europe* 19.6.87). Several surveys and estimations show that organic farming in Denmark is now an increasingly profitable production form and contains the potential to represent about one-tenth of the total agricultural sector within 10–15 years, both in terms of numbers of farmers and farmland area (*Agra Europe* 11.3.1988, 26.10.1990, 5.7.1991, 30.8.1991).

In the late 1980s/early 1990s, expectations of the future growth of organic farming and sales – especially of organic fruit and vegetables – had been noted in Spain. Similarly, in Spain in 1989, a regulating body, Consejo Regulador de la Agricultura Ecologica (CRAE), was established to register and authorise the use of a label for identifying food products produced according to specified organic production methods. Organic farming was seen mainly as containing a substantial potential for expansion in the less densely populated regions that still employed primitive farming methods, and as part of an export strategy, which the adoption of a common Community regulation of organic farming was expected to encourage even further (*Agra Europe* 27.1.1989, 22.2.1991).

Around the same time, a government plan was launched in the Netherlands to reduce the use of agri-chemicals in intensive agriculture and promote organic farming. Rather than pursuing an ever-increasing productivity in agriculture, the Minister of Agriculture (Gerrit Braks) held that intensive farming contributed to environmental depletion and, in turn, called for the agricultural sector to pay its share of the cost of environmental protection. Links were established between, on the one hand, the use of agri-chemicals and, on the other hand, problems of ground water pollution as well as acid rain. Furthermore, a link was established between the use of energy in glasshouses and problems of global warming (*Agra Europe* 26.5.1989). Although, the implementation costs of the Dutch plan were seen as a burden on farmers, by imposing increases in their costs and putting many out of business (*Agra Europe* 9.6.1989), a survey suggested that if the funding was made available, organic farming could develop to take up one-tenth of all agricultural land over a ten-year period (*Agra Europe* 1.6.1990).

Altogether, rather than being carried forward as a solution in the context of the CAP, organic farming in the UK seemed to be conceived of as a potential strategy for the development of a national niche production and market for organic produce. Moreover, organic farming in the UK was being linked to problems of increasing productivity and in the use of agrichemicals, that is, to intensive farming, but also seemed to be more closely associated with animal welfare issues than elsewhere (e.g. within the CAP, Denmark and the Netherlands). In Denmark, organic farming gained attention against the background of public concerns about water pollution caused by intensive agriculture and, here, organic farming was also seen as containing a potential in relation to the development of both home and export markets. In the Netherlands, organic farming was conceived of as part of a strategy to counteract the tendency towards ever-increasing agricultural productivity and the adverse effects of intensive agricultural production on the environment such as water pollution, acid rain and global warming. Organic farming in Spain is seen mainly as constituting a potential for less densely populated regions with primitive agricultural production methods and possible export markets. Finally, in France in the late1980s/early 1990s, organic farming was conceived of as part of a broader strategy aimed to give French agriculture a 'greener' image and to fulfil consumer demands for quality food products, while over the same period of time in Germany, a government report had shown the potential for greater profits for organic farmers than for those engaged in conventional farming (*Agra Europe* 4.5.1990, 22.2.1991a).

Regarding the exercise of entrepreneurship, in the UK, Denmark, Spain, the Netherlands and France organic farming began to receive attention as a subject for political regulation. In both the UK and Spain links were made between organic farming and problems within the CAP in the sense that in the UK organic farming was conceived of as a solution to problems imposed by the CAP, while in Spain, in the sense that a common Community regulation should be established in order to facilitate export markets. In the UK, Denmark, Spain and Germany organic farming was linked to certain consumer demands. Although organic farming in the UK, Denmark and the Netherlands was conceived of as a solution to certain environmental problems caused by intensive agricultural production, organic farming was not, however, immediately linked to the fulfilment of CAP objectives related to the protection of the environment. Finally, and particularly in Spain, organic farming was linked to issues relating to the maintenance of the countryside.

The Authorisation of Organic Farming (early 1990s)

In the early 1990s, the Agriculture Council adopted a Regulation (Council 1991; Reg. 2092/91), which laid down a Community definition of organic farming and a Council Regulation on agricultural production methods compatible with concerns for the environment, including organic production methods (Council 1992; Reg. 2078/92). Whereas the Community definition of organic farming adopted in June 1991 was first proposed by the Commission in late 1989, the Council Regulation concerned with support of agricultural production methods compatible with the environment was one of three 'accompanying measures' adopted in relation to the 1992 MacSharry reform (the other two being an early retirement programme and a farm forestry programme (Buller *et al.* 2000, p.3; Bruckmeier and Ehlert 2002, p.15).

From the 1991 Council Regulation it appeared that organic farming was to be seen in the context of a reorientation of the CAP and 'this type of production may contribute towards the attainment of a better balance between supply of, and demand for, agricultural products, the protection of the environment and the conservation of the countryside' (Council 1991; preamble). That is, the conception was held that organic production may be a partial solution to certain consumer demands of which the regulation aimed to encourage further through the adoption of a set of common production standards, and a common label for organic products. Organic farming was, on the other hand, also conceived of as balancing out the supply of agricultural produce, that is, it was further held that organic farming may also be a partial solution to problems of agricultural surplus production.

Additionally, it was considered that the development of a consumer driven market (a driving force the 1991 Council Regulation is meant to encourage further) for organic products and the expansion of organic productions methods may contribute to more environmentally friendly agriculture, this being a wished for development. A precondition for the working of the policy was that consumers would be willing to pay higher prices for organic products (Council 1991; preamble). It also appeared that 'the use of fertilizers and pesticides...*may* have detrimental effects on the environment or result in the presence of residues in agricultural produce' (Council 1991; preamble; own emphasis). This is as close as we get to the establishment of a causal relationship between intensive agriculture and environmental depletion at this point. However, one year later, any lingering doubts were removed and the link was authorised accordingly in the context of the MacSharry reform. Finally, the 1991 regulation also required the setting up of what became the Standing Committee for Organic Farming, which is made up by Member State representatives, chaired by a representative of the Commission with the purpose of assisting the Commission in an ongoing development of the rules for organic production (Council 1991; Art. 14).

In early 1992 the Commission adopted a Reflection Paper entitled 'The Development and Future of the CAP' elaborated by the DG for Agriculture, which is often referred to as the MacSharry Plan (Fennel 1997, p.169). The Reflection Paper reiterated the familiar cognitively instituted problems related to agricultural surplus production and budget pressures and, in this regard, the budget pressure exerted by the build up of intervention stocks. In addition, current problems in European agriculture were expressed as having to do with the unjust distribution of support among farmers and regions, environmental depletion, and a growing tension with external trading partners (Commission 1991, pp.1–3).

The causes of the problems identified were conceived of as both internal and external to the CAP. Developments outside of the Community, such as a decline in world market prices and a fall in the value of the US dollar, had negative effects on the problems central to the CAP with such external forces being regarded, at that time, of secondary importance to internal forces (Commission 1991, p.7). Internal sources of the problems of agriculture arose out of the basic mechanisms through which the CAP supported agricultural production, namely, the guaranteeing of certain prices for agricultural produce and direct product subsidies. These support mechanisms also contributed to problems related to environmental depletion since:

a system which links support to agriculture to amounts produced stimulates production growth and thus encourages intensification of production techniques. This development, if unchecked, leads to negative results. Where intensive production takes place nature is abused, water is polluted and the land [i]mpoverished. Where land is no longer cultivated because production is less dependent on surface area, abandonment and wilderness occur (Commission 1991, p.2).

The resolution of the current problems was envisaged by means of direct aid to farmers decoupled from product quantities, a restrictive price policy, a rural development policy and support for extensive farming methods (Commission 1991, pp.9–12). The support for extensive farming overall and organic farming as a sub-section, was introduced by a Council Regulation in mid-1992.

The 1992 Council Regulation came to a similar reasoning to that underpinning the regulation adopted a year earlier and it was claimed that organic production should be supported due to consumer demand for such products, the beneficial impact of organic farming on the environment, the protection of natural resources, the countryside and the landscape and, because environmental protection should be an integral part of the common agricultural policy (Council 1992; preamble; Art. 6). Moreover, 'both the Community and the Member States must increase their effort to educate farmers in, and inform them of, the introduction of agricultural and forestry production methods compatible with the environment, and in particular regarding the application of a code of good farming practice and organic farming' (Council 1992; preamble). That is, organic farming was placed on a par with 'good farming practice'.

While there had been some uncertainty in the 1991 Council Regulation all doubts had now been overcome as to the links between, first, intensive farming and environmental depletion and, second, between organic farming, on the one hand, and certain environmental benefits and the balancing of agricultural markets, on the other hand. For instance, the intention of the regulation was 'to promote...the use of farming practices which reduce the polluting effects of agriculture, a fact which also contributes, by reducing production, to an improved market balance' (Council 1992; Art.1) and 'the use of organic farming methods can help not only to reduce agricultural pollution but also to adapt a number of sectors to market requirements by encouraging less intensive production methods' (Council 1992; preamble).

By the end of this period, the problems relating to surplus production and the different problems relating to environmental depletion, which previously had been instituted cognitively and normatively respectively, had mutated in the sense that both types of problems were now being caused by intensive agriculture. That is, whereas intensive agriculture was conceived to be a legitimate problem within the CAP caused by technological progress and the modernisation of agriculture at the close of the previous period, intensive agriculture was now institutionalised as a source of problems that related to environmental depletion and surplus production rather than as problems per se.

Institutionalisations, Conflicts Over Meaning and Policy Entrepreneurship (1986–1992)

The development in problems, their sources, and solutions as articulated within CAP in the period from 1986 to 1992 may be summed up as shown in Table 6.1.

Taken together, the problems central to the CAP during this period represented continuity. Throughout the period from 1986 to 1992 the cognitively instituted problems of surplus production and budget pressure were thus central concerns to the CAP as it was in the previous period – with the qualification that the build up intervention stocks had become an intermediate factor (in the sense that they are products of surplus production and were exacerbating problems related to budget pressures). Similarly, although the decline of land and food quality was articulated but not institutionalised as an additional normative concern for the CAP, the normatively instituted problems of environmental depletion and the unjust distribution of support among farmers and regions represented, throughout the period from 1986 to 1992, institutionalised concerns within the CAP. In contrast to the end of the previous period, however, tensions with trading partners had, by the end of the current period, become a problem in itself and intensive farming was institutionalised as the source of surplus production and environmental depletion, rather than being regarded as a problem in itself. Moreover, the institutionalisation of normatively instituted problems alongside other cognitively instituted concerns by the end of the previous period, formed the bases for a number of alternative and, to some degree conflicting, conceptions as to the priority to be given to such problems, their the sources and solutions during the current period.

For instance, when priority is given to mainly cognitively instituted problems related to surplus production and budget pressures, the sources of such problems are a depressed international market and, in turn, a lack of demand on the Community markets, the implementation of the SEM as well as a decline of the value of the US dollar and the implementation of a trade distorting US food policy. However, when priority is given to more normatively instituted problems related to environmental depletion, rural exodus, pollution, and the decline of land and food quality, the sources of such problems are unclear.

Table 6.1: Problems and Solutions: The CAP and Organic Farming (1986–1992)

	CAP (end of previous period)	Alternatives	CAP (end of current period)
Problems	• Surplus production and budget pressures • Rural exodus, structural diversification, intensive agriculture, environmental depletion	• Surplus production, build up of intervention stocks and budget pressures (cognitively instituted problems) • Environmental depletion, rural exodus, pollution, decline of land and food quality (normatively instituted problems)	• Surplus production, build up of intervention stocks and budget pressure • Unjust distribution of support among farmers and regions • Environmental depletion • Tension with trading partners
Causes	• Modernisation of agriculture • Economic recession • Technological progress • Community enlargement • Choice of society in favour of a 'Green Europe'	i. Intensive agriculture, in turn, caused by technological progress Urbanisation, industrial development and transport issues ii. CAP principles: market unity, Community preference and financial solidarity iii. Link between modern agricultural production and environmental problems questioned. Scope of environmental problems questioned iv. Depressed international and, by association, Community markets. Implementation of the SEM, decline of value of US dollar and US food policy	• Guaranteed prices and direct product subsidies • Intensive farming • Decline in value of US dollar and world market prices
Type of Problem and Causality	• Cognitively instituted/normatively instituted problems • Multi-causal • Structural and thematic causes • Combination of causes inside and outside of the CAP with an emphasis on the latter	i. Mutation of normatively and cognitively instituted problems. Mono-causality. Structural and thematic causes. Causes essentially found outside of the CAP ii. Normatively instituted problems: mono-causality; structural and thematic causes; causes found within the principles of the CAP iii. Questioning normatively instituted problems and their causalities iv. Cognitively instituted problems: multi-causal; structural and thematic causes; causes found largely outside of the CAP	• Mutation of cognitively instituted/normatively instituted problems • Multi-causal • Structural and thematic causes • Combination of causes inside and outside of the CAP with emphasis on the former

Table 6.1: Problems and Solutions: The CAP and Organic Farming (1986–1992) (continued)

	CAP (end of previous period)	Alternatives	CAP (end of current period)
Solutions	• Agriculture as protector of environment • CAP dealing with agriculture as part of larger economy (Common Food Policy)	i. Extensive agricultural production ii. Environmentally informed CAP iii.Mechanism integrating environmental and agricultural concerns. Availability of Community solution questioned. Restrictive price policy questioned iv. Restrictive price policy, Co-responsibility, more flexible interventions, 'policy for quality'	• Direct aid and support for extensive farming • Restrictive price policy • Rural development policy
Organic Farming as a Solution	• Meeting demand from consumers (acceptability) • Labelling needed to ensure trust and legitimise premium prices (affordability) • Research needed on link between organic farming and food quality (not readily available)	i. and ii. Low input of energy and agri-chemicals improve ecological balance. Higher labour input to counteract rural exodus. Production of quality products to meet increasing consumer demand (acceptability). Lower yields hardly a problem! Potentially profitable niche production (affordability) iii. No subsidies. Should develop on market conditions and by demand of consumers. iv. None	• Meet consumer demand for protection of environment (affordability) • Contribute to CAP objective of environmental protection and maintenance of the countryside (acceptability)

On the one hand, links are articulated between such problems and intensive agriculture that, in turn, are caused by technological progress, urbanisation, industrial developments and transport. On the other hand, links are also articulated between current problems in agriculture and the principles of operation of the CAP, namely, the pursuit of market unity, preference for Community production and the financial solidarity of the CAP. Moreover, not only are the sources of current problems unclear, it is questioned whether a link exists between modern agricultural production and environmental depletion and, likewise, the scope of environment depletion is contested.

When priority is given to cognitively instituted problems related to surplus production and budget pressures, solutions envisaged have to do with a restrictive price policy, more flexible market interventions, and a more marginal and unspecified 'policy for quality'. At the same time, cognitively instituted problems and normatively instituted problems related to environmental depletion also mutate during the current period in the sense that intensive farming is conceived of as a central source of both problems. Solutions envisaged in this regard represent a CAP in support of extensive agricultural production and a more environmentally informed CAP. It has also been questioned whether a restrictive price policy is a viable mechanism to direct intensive agriculture towards more extensive production methods as well as whether Community solutions to environmental depletion are in fact available. Finally, on the one hand, organic farming is envisaged as a potential and partial solution to problems related to the extensive use of energy and agri-chemicals in intensive agriculture, rural exodus, consumer demands, food quality and certain environmental problems. On the other hand, it is also suggested that organic farming should, rather, develop on market conditions and as consumer demand for such products increases.

Whereas there is, from the beginning of the current period, some degree of conflict over the priority to the given to, respectively, cognitively and normatively instituted problems and, in particular, conflicts over the sources of such problems, these problems seem to have mutated in the sense that intensive farming has been institutionalised as the source of both, and conflicts are less marked by the end of the period. In addition, the sources and solutions to problems central to the CAP are specified towards the end of the current period.

Thus, the problems central to the CAP by the end of the previous period are all linked to broader structural developments inside and outside of the CAP (modernisation of agriculture, economic recession, technological progress, community enlargement, choice of society in favour of 'Green Europe'). The causes of the largely similar problems are by the end of the present period, however, specified international developments (a decline in value of the US dollar, falling world market prices) and specified regulatory mechanisms (guaranteed prices, direct product subsidies). In addition, problems caused by guaranteed prices, direct product subsidies, and also intensive farming, have available and acceptable solutions, namely, direct aid, a restrictive price policy, support for extensive agriculture and an active rural policy. In addition, organic farming has been institutionalised as an agricultural sector with certain distinguishable characteristics and as a subject for Community regulation, as a solution to certain consumer demands (whether for quality food products or environmental protection), and as contributing to the CAP in its endeavour to protect the environment and to maintain the countryside.

Against this background, a series of agents may also be identified as translators, as establishing forums for communication, and as carriers of concepts and conceptions, which was institutionalised during the current period and which links organic farming to the CAP. The concepts and conceptions institutionalised during the current period in this regard involves the launching of organic farming as an agricultural sector for political regulation and, essentially, community regulation as well as the establishment of links between, on the one hand, organic farming and, on the other hand, problems within the CAP, consumer demands, the fulfilment of CAP objectives related to environmental protection, the maintenance of the countryside and – under this broad umbrella – the conception that organic farming constitutes an employment opportunity in agriculture and potentially profitable niche production for farmers.

Table 6.2: Policy Entrepreneurship and the Institutionalisation of Organic Farming within the CAP (1986–1992)

Concepts and conceptions	Types of policy entrepreneurship		
	Translators	Establishing a forum for communication	Carriers
Organic farming as a sector for political regulation	• EP Com. Env.; Individual MEPs; EP Com. Agri. (alternative farming and not financially supported)	• EP Com. Env. (resolution and public hearing) • Individual MEPs (Parliamentary questions)	• EP; EP Com. Agri. • Commission; DG Env.; DG Agri. • Agriculture Council; UK; DK; ES; NL; FR
Linkage of organic farming to problems within the CAP	• EP Com. Env.; Individual MEPs	• EP Com. Env. (resolution and public hearing) • Individual MEPs (Parliamentary questions)	• EP; EP Com. Agri. • Commission; DG Env.; DG Agri. • Agriculture Council; UK; ES
Linkage of organic farming to consumer demands	• EP Com. Agri. (alternative farming); Individual MEPs	• Individual MEPs (Parliamentary questions)	• EP; EP Com. Agri. • Commission; DG Env.; DG Agri. • Agriculture Council; UK; DK; ES; DE
Linkage of organic farming to the fulfilment of CAP objective of environmental protection	• EP Com. Env.; Individual MEPs	• EP Com. Env. (resolution and public hearing) • Individual MEPs (parliamentary questions)	• EP; EP Com. Agri. • Commission; DG Env.; DG Agri. • Agriculture Council
Links organic farming to the maintenance of the countryside	• EP Com. Env.; EP Com. Agri. (alternative farming); Individual MEPs	• EP Com. Env. (resolution and public hearing) • Individual MEPs (Parliamentary questions)	• EP; EP Com. Agri. Commission; DG Env.; DG Agri. • Agriculture Council; ES

The exercise of policy entrepreneurship linking organic farming to the CAP in the period from 1986 to 1992 may thus be summed up as in Table 6.2. As carriers the EP, the EP Committee on Agriculture, the Commission, the DG for Agriculture, the DG for the Environment, the Council for Agriculture and a number of Member States have all contributed to the institutionalisation of the conceptions linking organic farming to the CAP as outlined above. However, the more vigorous type of entrepreneurship, which contributed to processes of translation and giving momentum to the institutionalisation of organic farming within the CAP, were exercised by the EP Committee on Agriculture and, particularly, individual MEPs, and the EP Committee on the Environment. It should be noted that – as a translator – the EP Committee on Agriculture was concerned with alternative agriculture rather than organic farming specifically. Finally, unlike the EP Committee on Agriculture, individual MEPs and the EP Committee on the Environment also give momentum to processes of institutionalisation by contributing to the establishment of forums for communication.

7

The Formation of a Policy Field: Organic Farming Within the CAP (1993–2005)

By the end of the period from 1986 to 1992 the central cognitively instituted problems within the CAP are related to agricultural surplus production, the build up of intervention stocks and budget pressures. In addition, by 1992, the CAP was also faced by a number of normatively instituted problems related to the unjust distribution of support among farmers and regions, environmental depletion and tensions with trading partners.

Alongside these concerns, it is proposed that the period from 1993 to 2005 may be characterised by the articulation and institutionalisation of a number of problems related to rural development, food safety and food quality. Although problems related to rural development are familiar to the CAP, such concerns are, during the current period, coordinated and put together in a 'second pillar' of the CAP dealing with rural development issues. Moreover, although, problems related to food quality and food safety (or consumer health) have been given voice within the CAP previously, it is proposed that it is not until the late 1990s that such problems have been elevated to institutionalised concerns of the CAP. Finally, while the previously cognitively instituted problems still supply matters for concern, it is proposed that during the current period such matters are, in fact, increasingly forming parts of analytically and normatively instituted problems within the CAP.

It is in this context that the formation of a policy field concerned with organic farming is taking place. It has been suggested that the CAP contains a series of commodity regimes each of which has 'its own distinctive problems and patterns of politics' (Grant 1997, p.102) and that commodity regimes evolving around, for instance, specific dairy and arable products are highly resistant to change and 'deeply embedded in the CAP's decision-making structures' (Grant 1997, p.102). While it may very well be so, it is proposed here that the period from 1993 to 2005 is characterised by the institutionalisation of what resembles a policy field concerned with organic farming within the auspices of the CAP. In that sense, the case pursued lends support to the suggestion that increasing emphasis has been put on various farming systems in the wake of the 1992 CAP reform – possibly at the expense of commodity regimes (Grant 1997, p.102).

More specifically, the formation of a policy field concerned with organic farming may be said to have taken place during the current period in the sense that, first, organic farming is continuously articulated and authorised as a potential and partial solution to certain and familiar problems as well as linked to a growing number of problems within the CAP. Second, while most – though not all – of the agents involved are familiar within the CAP, these agents are, on an increasingly regular basis, involved in disputes around an increasing number of issues related to organic farming. Third, that a policy field concerned with organic farming is in the making within the auspices of the CAP during the current period is probably best illustrated by the articulation of an increasing number of problems and conflicts related to the boundaries of organic farming vis-à-vis other policy fields. That is, problems are articulated with regards to, and conflicts appear over the boundaries for, what distinguishes organic farming from other concerns, what sorts of processes should be guiding this field, and which agents should be included and excluded. Importantly, it appears that solutions to such problems and conflicts are to be found within the 'new' or changed CAP.

In keeping with the definition of a policy field (see Chapter 3), it is thus proposed that organic farming – by 2005 – has been established within the CAP as (i) a system of problems and solutions, which links organic farming and the CAP, and within which disputes evolve around the nature of the links between organic farming and the CAP. This system and these disputes evolve (ii) among a set of agents representing the Commission, the Commission Services, the EP, Member States, research and various organised interests. Further, these agents (iii) operate according to the consultation procedure, interact at successive conferences and, importantly, among the agents it is commonly agreed solutions to existing disputes should be pursued in the context of the CAP.[15] Finally, (iv) it is – to some degree – possible to distinguish this policy field from other fields of concerns, agents and processes.

---

[15] To this list could be added the 'day-to-day' work taking place in the Standing Committee on Organic Farming, which was established in mid-1991.

It is further proposed that the formation of a policy field concern with organic farming within the CAP during the period 1993 to 2005 is given momentum through a series of conflicts over meaning, the exercise of policy entrepreneurship and the existence of a widely recognised crisis. Conflicts over meaning are found, for instance, within the Commission, the EP and at successive conferences. Furthermore, a wide range of agents gives momentum to the formation of a policy field concerned within organic farming within the CAP. However, it is proposed that the EP Committee on the Environment, and (much more particularly so) the DG for Agriculture have exercised the more vigorous type of policy entrepreneurship by contributing to the translation of organic farming within the CAP during the period under investigation. Finally, it is proposed that the 'BSE crisis' was a conducive condition for the formation of a policy field concerned with organic farming within the CAP from 1996 onwards. The following will first consider the discursive and institutional developments within the CAP towards the end of the period under investigation and, subsequently, the discursive and institutional developments marking the formation of a policy field concerned with organic farming within the auspices of the CAP. A total number of 188 empirical documents and articles are consulted.

Ideas Within the CAP (Towards the End of the Period)

A reform of the CAP was undergoing preparation during the latter part of the 1990s and in March 1998 the Commission presented a set of 'Proposals for Council Regulations (EC) concerning the reform of the common agricultural policy (Commission 1998)'. The DG for Agriculture was primarily responsible for the elaboration of these proposals but an additional number of DG's were associated including, among others, the DG for Competition, the DG for Regional Policy and Cohesion and the DG for the Environment. The ideas articulated in the 1998 Commission document were eventually endorsed by the Council at the Berlin Summit on Agenda 2000 in March 1999 and subsequently referred to as the 1999 CAP reform. Although conflicts do evolve around the particularities of the various elements suggested in the Commission document from 1998, the problem figures outlined were carried through into the 1999 reform with the qualification that the then recently implemented Amsterdam Treaty was considered to be further enforcing the objective to integrate environmental concerns into all areas of EU policies (Commission 1999a, pp.1–3). Accordingly, the problems to be addressed within the CAP towards the end of current period are made up by largely familiar elements, yet some of these are instituted slightly differently from earlier periods. Surplus production, the build up of intervention stocks and related budget pressures are still matters of concern but now they are articulated as the potential consequences – as is rising unemployment – of a lack of international competitiveness in European agriculture. That is, a central problem within the CAP is whether European agriculture will be able to profit from more liberalised international agricultural markets (Commission 1998, p.2). This problem may be characterised as analytically instituted in the sense that it is based on a coupling between, on the one hand, a predicted further liberalisation of international trade with agricultural products and potentially profitable international markets and, on the other hand, actual developments within the CAP, which has established an experience of a relationship between high price policies and insufficient competitiveness in the world market. The sources of this problematic are mono-causal – in the sense that they are clear and simple – structural and thematic and found both outside and inside the CAP. Problems of liberalised international agricultural trade are, thus, partly related to a predicted strong growth in demand and prices in the world markets for agricultural produce, and partly related to the still high price policies within the CAP (Commission 1998, p.2, 5). The solutions suggested in this regard were cuts in the guaranteed prices accompanied with an increase in direct income support to farmers (Commission 1998, p.5) and, hence, represented a continuation of solutions institutionalised within the CAP by the end of the previous period.

At this point in time, a second problem central to the CAP is related to 'how agricultural policy is devised and managed' (Commission 1998, p.3). Whereas the CAP was established to manage agricultural policies among the original six Member States, it was argued that the management of the CAP needs to be adjusted to accommodate not only the current 15 Member States but also to take into account the accession of the new Member States from central and eastern Europe. The enlargement is predicted to enhance the complexity and bureaucracy of the CAP and the diversity of agriculture in the EU. Although diversity in natural conditions, production methods, income levels and competitiveness is a quality of agriculture in the EU, increased diversity enforces the need to take into consideration the requirements of particular sectors and local conditions (Commission 1998, p.3). The second problem within the CAP may also be characterised as analytically instituted in the sense that it is based on a coupling between, on the one hand, the predicted enlargement of the EU to the east and, on the other hand, a number of actual enlargements throughout the development of the CAP, which has established an experience of a relationship between Community enlargements and structural diversity, and administrative complexity.

The source of the problems (i.e. the enlargement to the east) relating to the management of the CAP is mono-causal, structural, found outside of the CAP but also thematic since enlargements have been a frequently reoccurring concern throughout the development of the CAP. The solutions envisaged are in general aimed at the establishment of a 'new balance between common management and increased decentralisation'. More specifically, it is suggested that 'national envelopes' should be established, which entail that resources are allocated from the CAP budget to Member States according to national agricultural production size which, in turn, are distributed by Member States according to their particular objectives vis-à-vis national agriculture (Commission 1998, p.5).

The third problem for the CAP to deal with has to do with its degree of legitimacy among the wider public. While the unjust distribution of support among farmers and regions is still a concern for the CAP, the unjust distribution of support is now conceived, on the one hand, as 'having negative effects on regional development planning and the rural community, which has suffered badly from the decline in agricultural activity in many regions'. On the other hand, 'other regions have seen the development of excessively intensive farming practices which are having often a serious impact in terms of the environment and animal diseases' (Commission 1998, p.3). Although there is not established a direct causal relation between the support for already better-off regions and problems related to the environment and animal health, together '[a]ll these factors combine to create a bad image of the CAP in the minds of the public' (Commission 1998, p.3). From the 1998 Commission document, it also appears that '[m]aking the CAP more acceptable to the citizen in the street, to the consumer, is one of our primary tasks in the years ahead' (Commission 1998, p.3). Moreover, '[a]n agriculture which pollutes, which contributes inadequately to spatial development and protection of the environment, and which, because of its undesirable practices, must take its share of responsibility in the spread of animal diseases, has no chance of long-term survival and cannot justify what it is costing' (Commission 1998, p.3). The image problem of the CAP may be characterised as normatively instituted in the sense that it is based on a coupling between, on the one hand, an actual but also predicted further inadequacy of the CAP to resolve current problems in agriculture and, on the other hand, an ideal holding that the CAP needs to obtain its legitimacy in the wider public.

The causes of problems related to lack of legitimacy or acceptance of the CAP among the wider public are thus multiple, structural and thematic in nature (distorted distribution of support among farmers and regions, the decline of activities in certain rural areas, intensive agriculture, environment depletion and animal health issues) and are found within the CAP in the sense that the policy so far has been inadequate. Solutions envisaged to improve the image of the CAP among the wider public are, for instance, a simplification of the CAP and, in general, 'it is vital to deal with various inequalities and abuses which seriously harm the image of the CAP' (Commission 1998, p.6). Great hope is attached to the introduction of a 'second pillar' of the CAP dealing with rural development which, in essence, put together existing measures directed towards environmental protection and structural improvements in rural areas (Commission 1998, p.6, 134).

Most recently, proposals for a number of Council regulations have been elaborated by the DG for Agriculture, adopted by the Commission, endorsed by the Council and subsequently referred to as the 2003 CAP reform (Council 2003). The proposals suggested by the Commission and subsequently endorsed by the Council largely reiterate problems institutionalised in the context of the

1999 CAP reform. This appears in the explanatory statement to the Commission proposals. The problems the CAP needs to address thus have to do with the competitiveness of European agriculture, the lack of responsiveness to market developments, broader societal expectations directed towards the CAP and issues related to rural development. Additionally, both the lack of responsiveness of agricultural production to market developments and concerns related to rural development is linked to CAP objectives of environmental protection, food quality, food safety and animal welfare concerns (Commission 2003, p.3). A central cause of current problems is the preferred support mechanisms of the CAP that are conceived as establishing an incentive system, which has negative effects on the protection of the environment and, in general, counteract moves towards more sustainable agricultural production (Commission 2003, p.5).

While the solutions proposed are still conceived in terms of preparing the CAP for the forthcoming enlargement to the east, it is also envisaged they will strengthen the position of the EU in the context of the forthcoming negotiations in the WTO (Commission 2003, p.5). The preferred and familiar solution of decoupling agricultural support from yields and enhanced direct support to farmers is conceived to address problems regarding the competitiveness of European agriculture, the lack of responsiveness to market developments, and rural development. However, the enhancement of direct support to farmers is conditional on agricultural production methods satisfying certain requirements in terms of protection of the environment, food safety, food quality and animal welfare and, in cases where these conditional requirements are breached, sanctions will be enforced (Commission 2003, p.3, 10). While the 2003 CAP reform, in the main, seems to represent a continuation of the already institutionalised problems and solutions, concerns with food safety, food quality and animal welfare is reinforced (possibly at the expense of environmental concerns) by being linked to both market developments and rural development and by being connected with legal sanctions.

Organic Farming Within the CAP (1993–2005)

Although organic farming is not an issue for the Commission preparatory document for a Rural Development policy and seems almost to be lost in the process, organic farming reappears in the final Council Regulation on rural development, this having been adopted as part of the 1999 CAP reform. The Council here endorsed that 'demand from consumers for organically-produced agricultural products and foodstuff is increasing…a new market for agricultural products is thus being created…[and] organic agriculture improves the sustainability of farming activities and thus contributes to the general aims of this Regulation [of rural development]' (Council 1999b, p.83). Moreover, an additional 'chapter' on food quality was introduced by way of the 2003 CAP reform into the rural development policy which, among other things, officially acknowledges organic farming as a production method producing quality food products (Commission 2003, p.68).

To be sure, organic farming is but one among a wide series of problems and solutions in the context of the rural development policy. However – and concurrently with the elaboration, adoption and implementation of the 1999 and the 2003 CAP reforms – the formation of a policy field within the CAP seems to take place around the issue of organic farming.

Conflicts Over Boundaries and Policy Entrepreneurship Within the Commission (mid-1990s)

When organic farming was first articulated as a Community concern in the context of the emerging EC environmental policy in the mid-1970s it was, among other things, linked to a rising level of attention given to food quality among consumers. Likewise, in the period from 1978 to 1985 organic farming was among people involved in alternative agriculture and – towards the end of the period – also within the CAP linked to consumer demands for high quality food products: this came with the significant qualification that the Commission and the Council both issued calls for research into the relationship between production methods and food quality. During the subsequent period from 1986 to 1992, the link between consumer demands for quality food products and organic farming was voiced within the EP, however, within the Commission and the Commission Services there was still some uncertainty as to the existence of such a link. By the end of the period, the conception that organic food products are not of superior quality vis-à-vis conventional agricultural produce was endorsed by a Council Regulation. In summary, and in various contexts, organic farming was articulated as an available response to consumer concerns with food quality, however, at no point has this conception become widely accepted.

In 1994, however, the DG for Information published a report dealing with the theme of organic farming in the EU in the series 'Green Europe', which is aimed at a broader public and which was noted by *Agra Europe* (10.2.1995). The report was executed by people within the DG for Agriculture against the background that '[t]he conditions for wider recognition and development of organic farming have been fostered in recent years by the development and adjustment of the common agricultural policy, and more generally by new ideas about the future of the countryside, and by emerging political awareness of environmental issues' (Commission 1994, p.1). Two main problems are identified in the report. One problem is concerned with 'less competitive agricultural areas' in the Community and another with the nature of conventional agricultural products. On the one hand, certain agricultural areas are seen to be facing problems of rural exodus, inefficient farm structures and natural handicaps such as less fertile soil, lack of water or mountainous areas disfavouring industrialised agricultural production. On the other hand, conventional agricultural products are seen as increasingly uniform and produced by methods involving intensive use of agri-chemicals and fertilisers. Such production methods and products have come to constitute a problem as consumers have redefined their conception of food quality. The notion of quality attached to conventional production and products have been concerned with 'standardisation and homogeneity': increasingly, however, consumers are turning away from this notion of quality and instead link quality to more 'natural' agricultural products (Commission 1994, p.1).

The problems relating to less competitive agricultural areas and the nature of conventional agricultural products may both be characterised as normatively instituted. The first in the sense that it is based on a coupling between, on the one hand, an actual but also predicted further rural exodus and inefficiency of farm structures and, on the other hand, an ideal conception holding that the CAP should contribute to the maintenance of the countryside especially in regard to the protection of less competitive agricultural areas. The second problem is normatively instituted in the sense that it is based on a coupling between, on the one hand, an actual but also predicted further uniformity of conventional agricultural produce and, on the other hand, an ideal conception holding that the CAP should contribute to the fulfilment of rising consumer expectations of the availability of quality food products. Moreover, the sources of the problems identified are simple, structural and thematic in nature (rural exodus, inefficient farm structures, natural handicaps/ intensive farming, and consumers' expectations of food quality).

However, rather than attributing blame for the problems identified, in particular, blaming the CAP, the recently reformed CAP is articulated as the forum where solutions should be pursued. Although it is not the only possible solution and its potential is modest in terms of its predicted share of the total agricultural markets by the year 2000 (2.5%), organic farming is seen as addressing both of the problems identified. Conversion to organic production methods is thus conceived as containing the potential to absorb surplus labour since it is labour intensive. Further, it is seen as supplying viable and profitable production methods, particularly in agricultural areas that are otherwise characterised by inefficient farming structures and natural handicaps, because of its low input production methods and the ability of organic produce to command higher prices. Additionally, organic produce is free from artificial chemical residues, contributes to the diversification of agricultural food products, and uses production methods favourable to the protection and conservation of the environment. Finally, through the eyes of the DG for Agriculture, consumers perceive organic food products as natural products, that is, high quality products (Commission 1994, pp.1–3, 24–25).

In 1996, an EU financed conference on organic farming was convened in a cooperation between the DG for Agriculture and the European Training and

Development Centre for Farming and Rural Life (CEPFAR), which has, among others, COPA and COGECA (General Committee for Agricultural Cooperation in the European Union) included in its membership (CEPFAR 1996). At this conference, organic farming was similarly identified by the DG for Agriculture as a response to two broader societal developments. That is, organic farming was articulated as a response to, on the one hand, a growing societal concern for the protection of the countryside and the environment and, on the other hand, a consumer demand for quality food products (DG for Agriculture 1996, p.5). The DG for Agriculture had already in 1994 pointed to problems within organic farming relating to the inefficient marketing of organic produce vis-à-vis consumer expectations of the presentation and continuous supply of such products (Commission 1994, pp.24–25). A marked feature of the 1996 conference was, however, that organic farming was articulated as being faced by a number of problems in its further development within the CAP. Hence, organic farming was seen to be facing problems that related to the forthcoming expansion of the EU regulation on the area including: the development of an EU logo for organic food products, the use of GMO's (Genetically Modified Organism) in organic food production and the carrying of the EU regulation into an international context, that is, the development and setting up of uniform international standards for organic production (DG for Agriculture 1996, p.11).

In summary, apart from confirming the by now institutionalised links between, on the one hand, organic farming and, on the other hand, CAP objectives regarding the protection of the environment and the maintenance of the countryside (rural exodus, less competitive rural areas), organic farming was from the beginning of the current period articulated as a potential and partial solution to problems within agriculture relating to the quality of conventional food products. This differs from expressions of uncertainty on the relationship between organic farming and food quality both during the previous period within the Commission and the Commission Services and, during the period up to 1985, within the Commission, the Commission Services and the Council. Likewise, it is in conflict with the conception that organic food products were not to be considered of superior quality vis-à-vis conventional food products as authorised by a Council Regulation towards the end of the period running from 1986 to 1992.

Regarding the exercise of policy entrepreneurship, the DG for Agriculture contributed to the translation of some of the conceptions that was institutionalised during current period. That is, the DG for Agriculture contributed to the translation of the conceptions, first, that intensive agricultural production is the source of problems related to food quality and, second, that organic farming constitutes a potential and partial solution to such problems. Third, the DG for Agriculture contributed to the translation of a number of problems facing organic farming in relation to its further development and identified the CAP as the forum where such problems should be resolved. While such conceptions have largely been articulated as alternatives to institutionalised concerns within the CAP in previous periods, it is unclear in which context problems facing organic farming are made available for translation. It was indicated in a speech by the president of IFOAM held at the 1996 conference (IFOAM 1996), however, that concerns with the boundaries of organic farming vis-à-vis other agricultural production methods are made available by people representing organised interests within organic farming (see below). Finally, the DG for Agriculture – together with CEPFAR – also exercises policy entrepreneurship through the establishment of a forum for communication in the form of a conference on organic farming in the EU.

The BSE Crisis (1996 onwards)

In 1996, concerns with BSE (Bovine Spongiform Encephalopathy) was turned into an EU-wide crisis after research had shown that the consumption of meat from BSE infected livestock can be fatal for humans and the Commission adopted a ban on the exports of beef from the UK, where the outbreak of BSE was particularly severe and received a great deal of attention (Grant 1997, pp.123–124; Roederer-Rynning 2003, p.123). It has been argued that the:

consequences of the BSE crisis included the diversion of the attention of EU decision-makers from other pressing problems; new tensions between member states, making consensus formation more difficult; new budgetary problems that are difficult to resolve, given internal and external constraints; a severe structural surplus problem in a key product sector; and an undermining of public confidence in modern systems of farming (Grant 1997, p.129).

If by the latter is meant 'modernised and industrialised' agricultural production opposed to farming methods 'closer to natural processes', the damage made by the BSE crisis on public confidence in modern food production seems to be a conducive condition for the institutionalisation of organic farming as a solution to food safety and quality problems from 1996 onwards.

So far, the common EU regulation aimed at the organic farming sector had not covered organic livestock production although a deadline had been set up for the Commission to elaborate a proposal to extend EU regulations to also apply to this part of the sector 'within a suitable period' (Council 1991). In July 1996, the Commission adopted a proposal for a Council Regulation on the matter (Commission 1996), which was elaborated under the jurisdiction of the DG for Agriculture in association with the DG for the Environment, the DG for Industry and the DG for Consumer and Health Policy. Although the Commission has argued that the proposal was delayed due to the complexity of the matter (*Agra Europe* 26.11.1993) and a Council Regulation was not adopted until mid-1999 (Council 1999), the proposal was timely as it appeared in the context of the 'BSE crisis' peaking in 1996 and again in late 2000/early 2001.

Organic farming is thus articulated as a potential and partial response to widespread concerns with the livestock disease BSE and food safety concerns in the mid-1990s. The 'BSE scare' in the UK is as early as in 1990 seen as forming the basis of rising consumer and supermarket demands, particularly for organic beef (*Agra Europe* 24.8.1990). Later, throughout 1996, the market for organic produce in the UK – not only on organic meat production – was seen as expanding against a background of an increasingly severe BSE crisis (*Financial Times* 6.7.1996, 19.7.1996, 20.7.1996). It was also around this time BSE became a problem to be dealt with in the context of the CAP which, among other things, formed the basis of the articulation and institutionalisation of organic farming as a solution to food safety problems. For instance, the Commission proposal for a Council Regulation of organic livestock production in mid-1996 was articulated as a response to problems with environmental depletion and animal welfare. It had been argued that the free movement of organic meat production within the EU must be ensured, that such production had a 'huge market potential' and was 'matching perfectly the aims of the CAP reforms of 1992' (*Agra Europe* 2.8.1996, 16.5.1997). In addition, however, the proposal was also timely since '[t]he quest to improve consumer information and confidence is an issue, the Commission observes, that has become highly important since the BSE crisis erupted' (*Agra Europe* 2.8.1996). Likewise, within the EP, the introduction of a common regulation of organic livestock production was conceived to be – among other things – a response to the BSE crisis and a means of promoting food safety. This appears through an EP report elaborated by the EP Committee on Agriculture[16] and subsequently adopted by the EP. The report, which contains 100 EP amendments to a regulation proposed by Commission, thus argues in the explanatory statement that the amendments suggested 'are, in late 1996, particularly significant for the European Parliament's political role in the **context of the current BSE crisis**' (European Parliament 1997, p.58; original emphasis). Moreover:

[t]his crisis calls for a thorough reworking of the quality control systems for food products, in particular livestock products and the regulations on organic livestock farming which remain very diverse depending on the Member State, and for a new consumer protection policy in the European Union and consumer guarantees in the Union's external markets (European Parliament 1997, p.58).

The rapporteur in charge of the drawing up the EP report (Christine Barthet-Mayer) for the EP Committee on Agriculture also argued in *Agra Europe* (16.5.1997) that this regulatory framework had become particularly relevant in

---

[16] The full title of this EP Committee is now the Committee on Agriculture and Rural Development but it is essentially carrying on the responsibilities related to the CAP that were previously carried by EP Committee on Agriculture, Fisheries, and Food.

the context of the BSE crisis. Furthermore, an attached opinion to the EP report by the EP Committee on the Environment begins by stating that:

[i]n the meat sector consumers, much more than in the past, are interested not only in the cost but also the quality of the products and, in particular, the production methods used. This interest on the part of European consumers has increased over the last few years, especially after the hormone scandal and the events surrounding BSE, which caused a great deal of concern among the general public (European Parliament 1997; Attached opinion by the EP Committee on the Environment, p.62).

Finally, that organic farming within the EP is broadly conceived to be an available and acceptable, albeit partial, response to the BSE crisis appears at two EP debates in 1997 and 1999. Eight out of twenty, and three out of nine speeches by MEPs in 1997 and 1999 respectively thus referred to the BSE crisis when arguing in favour of the expansion of the EU regulation of the organic farming sector (European Parliament 1997a, 1999).

The second peak in the BSE crisis in late 2000/early 2001, among other things, formed the background for a replacement of the German Minister for Agriculture, and a Cabinet reshuffle, but also the articulation of a turn in German agricultural policy in favour of organic farming (see below). On a Community level, the Commissioner for Agriculture (Franz Fischler), for instance, launched a Commission '7-point plan' in early 2001 as a reaction to the BSE crisis and its impact on the market for veal and beef (Commission (2001a). The first point of this plan bore the heading: 'Boosting organic farming' and stated that 'the BSE crisis demonstrates the need for a return to farming methods that are more in tune with the environment' (Commission 2001a). In fact, at the time, the Commissioner for Agriculture argued that '[a]voiding the beef mountains looming on Europe's horizon is in the interests of consumers, taxpayers and farmers alike. Immediate action to curb beef production by boosting less intensive and organic production is the only way forward' (Commission 2001a).

In all, it appears that within the Commission and the EP organic farming is, among other things, conceived of as a response to food safety issues and, in particular, to the BSE crisis that first peaked in 1996. Additionally, although the Council did not immediately link the adoption of a Council regulation on organic livestock production in mid-1999 to the BSE crisis, the Council did link organic farming to consumer concerns with food safety (Council 1999c). Finally, whereas the BSE crisis peaking in 1996 is conducive to the articulation of a link between food safety issues and organic farming, this link was carried on beyond this point in time being reinforced in late 2000/early 2001 by a renewed attention given to the outbreak of BSE.

Conflicts Over Boundaries Within the EP and the Commission
(late 1990s)

Organic farming in the late 1990s is conceived as faced by two central problems, which have to do with the use of GMO's in organic production and whether organic farming is a solution to certain human health concerns or not. The latter also appeared as a conflict over which legal decision-making procedure should guide formal decision-making on matters related to the regulation of the organic farming sector.

The EP Committee on Agriculture report from 1997 (see above) confirms the institutionalised links between, on the one hand, organic farming and, on the other hand, CAP objectives regarding the maintenance of the countryside and the protection of the environment (European Parliament 1997, p.57). However, the EP also found that 'genetically modified organisms...and products derived therefrom are not compatible with organic methods and principles; [and]...in order to maintain consumer confidence in organic production, genetically modified organisms, parts thereof and products derived therefrom must not be used in products labelled as organic' (European Parliament 1997, p.6). That is, the use of biotechnology in organic farming was not seen as compatible with consumer expectations related to organic food products. This conception is reiterated throughout the EP report and, likewise, it is endorsed by eight out of twenty MEPs in speeches made at an EP debate held in relation to the EP Committee on Agriculture report on organic livestock production (the group of eight MEPs mentioned here is a different group from the one referred to above) (European Parliament 1997a).

The Commission had not previously argued in favour of either the inclusion or exclusion of GMO's in organic farming. However, at the 1997 EP debate the Commissioner for Agriculture (Franz Fischler) recognised that the EP 'strongly emphasizes the need to ban the use of genetically modified organisms...in organic farming'. Additionally, '[t]his view is shared by associations of organic producers and the majority of Member States. The Commission therefore accepts that the use of genetically modified organisms...does not reflect the current expectations of consumers regarding organically produced agricultural products and foods' (Commissioner for Agriculture in European Parliament 1997a; see also *Agra Europe* 16.5.1997). After the elaboration of a report from an EU group of national experts on the use of biotechnology in organic production, and in spite of the belief that further tensions with the US would be sparked off around the issue of biotechnology and food production, it was unanimously agreed within the Council that all biotechnology-derived products should be banned from organic farming (*Agra Europe* 31.7.1998). It was also this conception – that biotechnology is not compatible with consumer expectations of organic produce – that was authorised by the Council and carried into the final regulation in 1999 (Council 1999).

A conflict was also evident over whether human health concerns should be included or excluded from the still evolving policy field concerned with organic farming. In the explanatory statement of the 1997 EP report, organic food production was thus expressed as a response to human health concerns in food production and 'organic farming is the agricultural sector with the closest link to consumer and public health policies in the European Union' (European Parliament 1997, p.58). The EP debate in relation to the EP Committee on Agriculture report also drew on this conception. For instance, it was argued that 'traditional farming has had a major adverse impact on the environment and also on consumer health'. In this regard, it is considered pivotal 'to accept that the Common Agricultural Policy will gradually have to be realigned towards more ecological production of plant and animal products. We believe, then, that organic farming is pointing the way that must be followed by farming as a whole' (Draftsman of opinion of the EP Committee on the Environment in European Parliament 1997a). Moreover, the conception that organic farming is a potential and partial solution to human heath concerns in food production was articulated by six out of twenty MEPs in speeches made during the 1997 EP debate (European Parliament 1997a). Yet it is also a conception, which was not endorsed by the Council Regulation adopted in mid-1999 (Council 1999) and the Commissioner for Agriculture (Franz Fischler) stated that the EU regulation of the organic farming sector 'is primarily concerned with regulating and promoting organic production methods, and thus has no health policy objectives' (Commissioner for Agriculture in European Parliament 1997a).

On the one hand, it is thus widely accepted that the use of biotechnology is not compatible with consumer expectations directed towards organic food products and – within the EP – it is argued that consumer confidence in such products depends on the ban of the use of GMO's in organic farming. On the other hand, neither the Commission nor the Council accept organic farming as related to human health concerns in food production. Moreover, opposed to, for instance, the uncertainty surrounding the link between organic farming and food quality that gave rise to a call for research into the matter in the mid-1980s, the link between organic farming and human health concerns in food production was rejected by the Commission on the grounds of relevance and procedural considerations. The Commission thus upheld that organic farming should be dealt with under the auspices of the CAP according to the formal decision-making procedure, which entail that the EP is consulted on such matters (Commissioner for Agriculture in European Parliament 1997a). Against this, it has been argued within the EP that organic farming is a matter of human health and consumer protection which – had it been endorsed – had the potential to endow the EP with enhanced formal decision-making powers along the lines of the co-decision procedure (European Parliament 1997; 1997a).[17] In other words,

---

[17] See for instance Nugent (1999, pp.207–209) for an overview and discussion of the formal powers attributed to the EP in the context of the consultation and co-decision procedures respectively.

conflicts appeared over the issues to be included and excluded from the field concerned with organic farming and over the legal procedures for guiding this field.

Conflicts Over Boundaries and the CAP as a Solution –
the Baden Conference (1999)

Towards the end of current period, the documents with the widest endorsement among the agents involved in organic farming within the auspices of the CAP were: the 'Conference Summary Statement' from 1999, the 'Copenhagen Declaration' from 2001, a Commission Staff Working Paper entitled 'Analysis of the possibility of a European Action plan for organic food and farming' that was issued in late 2002 and a European Action Plan for Organic Food and Farming from 2004.

The 1999 Conference Summary Statement was elaborated and presented at the end of a conference under the title 'Organic Farming in the European Union – Perspective for the 21st Century'. The conference, which was held in Baden, Austria in May 1999, had about 180 participants from across the EU and accession Member States. The participants included the Commissioner for Agriculture (Franz Fischler), the Commissioner for the Environment (Ritt Bjerregaard), the Austrian Federal Minister for Agriculture and Forestry (Wilhelm Molterer), and the Austrian Federal Minister for the Environment (Martin Bartenstein). Other participants included representatives from EU Member States, and EU accession countries as well as Commission officials, researchers, representatives of various environmental interest organisations, and representatives from both conventional farmers' and organic farmers' interests and market organisations (Commission 1999, pp.195–200). A number of familiar links between, on the one hand, organic farming and, on the other hand, certain environmental benefits, rural development, and consumer demands were reiterated reflecting the firm institutionalisation of these links at this point in time. Yet, it also appears that organic farming was contributing to animal welfare (Conference Summary Statement in Commission 1999). To varying degrees, speeches made by the Commissioner for Agriculture, the Commissioner for the Environment and people within the DG for Agriculture, proclaimed organic farming as a potential and partial solution to problems related to food safety, consumer demands for quality food products, economic activities in rural areas, and animal welfare. Moreover, organic farming was conceived of as a potential and partial solution to a series of environmental problems related to biodiversity, the quality of soil, pollution of ground water, the use of local and renewable resources and the protection of certain valued landscapes (Commissioner for Agriculture; Commissioner for Environment; 2 x DG Agriculture in Commission 1999).

A marked feature is, however, that the CAP, rather than being identified as the source of current problems, was endorsed as the forum for the resolution of problems facing organic farming (Conference Summary Statement in Commission 1999). In a series of speeches made at the conference this point was qualified and it appears that a changed CAP and further EU regulation was seen as the way forward for the development of organic farming in Europe. Although the advance of organic farming in Member States, the development of national regulatory frameworks and national action plans have been the first steps in the development of the organic farming sector, further action at EU level is now timely. Moreover, not only is EU action timely, but also the time must end where uncoordinated national solutions are pursued (Commissioner for Agriculture; Commissioner for Environment; 2 x DG Agriculture; Austrian Minister for Agriculture; Danish Ministry for Agriculture in Commission 1999).

A second marked feature of the 1999 conference is the lack of 'blaming', that is, the sources of the problems to which organic farming is conceived of as providing a solution were rarely articulated. However, one such rare example appeared in a speech by the Commissioner for Agriculture. Accordingly, one source of problems relating to food quality and safety and the competitiveness of European agriculture was identified as external to the CAP. More specifically, it was considered that hormone-treated meat and products and their importation into the EU would compromise quality and food safety concerns and the competitiveness of European agriculture (Commissioner for Agriculture in Commission 1999).

On the one hand, during the current period, organic farming is increasingly conceived of as being faced with a number of problems and conflicts related to the boundaries of organic farming vis-à-vis other concerns and, on the other hand, the 'new' CAP is increasingly conceived of as the forum where such conflicts should be fought and problems resolved. The problems facing organic farming include the increasingly manifest competition from non-organic but high quality food products and, for instance, it has been stated that:

European farmers are not the only farmers who produce organic food or produce according to high standards. Quality and safety are essential elements of the European model of agriculture, but high standards will not prevent us from competition and imports. It is a misunderstanding, that our European level is so much higher than in other countries, that no food would comply with our requirements and would be allowed to enter into the common market (Commissioner for Agriculture in Commission 1999).

Organic farming is, thus, not only subject to competition on the production of high quality food products from European agriculture in general but also from quality food production outside of the EU. In addition, it is reiterated that there exists a need to maintain the EU-wide prohibition of the use of GMO's in organic farming in order for organic products to be in line with consumer expectations (Conference Summary Statement in Commission 1999). Although a ban on the use of GMO's in organic food production was not formally adopted until a few weeks later (Council 1999a), the Commissioner for the Environment linked the need for a ban to the existence of an EU level commitment to uphold the 'integrity of organic farming' (Commissioner for Environment in Commission 1999). That is, organic farming is, on the one hand, faced by problems related to consumer expectations about the nature of certain biotechnologies and their use in organic production and, on the other hand, the CAP supplies the solution in the form of an EU ban and, hence, secures the continuous 'integrity of organic farming'. Dissimilar national support schemes and developments in organic farming is also a source of concern, which builds up to the search for solutions at the EU level, that is, within the CAP. Thus:

the conference recognised the potential of organic farming to grow from 2% currently to 5–10% of EU agriculture by 2005, but noted that the subsidiarity principle and resource constraints could lead to substantial differences in the level of support for organic farming between member states and that these issues would need to be kept under review (Conference Summary Statement in Commission 1999).

Finally, it even appears that the environmental benefits of a few and certain types of organic production are questionable. In general, the conference concluded that EU regulation should have the objective 'to maintain a high level of credibility and integrity of organic production, by implementing high production standards and strict inspection requirements, and by effectively communicating the objectives and principles of organic farming' (Conference Summary Statement in Commission 1999).

Policy Entrepreneurship Within the EP and the Commission (late 1990s)

Regarding the exercise of policy entrepreneurship, earlier in current period the DG for Agriculture had given voice to a concern with, and the need for, examining the relationship between GMO's and organic production (DG for Agriculture 1996). However, it is within the EP this relationship became clarified and the conception articulated that the use of biotechnology in organic farming is not compatible with consumer expectations directed towards organic food products. More specifically, on the one hand, it was the EP Committee on Agriculture, which prepared the EP report containing, among other things, the EP amendments in favour of a prohibition of the use of GMO's in organic farming and, hence, this Committee endorsed that GMO's should be excluded from organic farming. On the other hand, however, it was the EP Committee on the Environment, which instigated the exclusion of GMO's in organic production in an attached opinion to the report (Attached Opinion in European Parliament 1997). Hence, the EP Committee on the Environment exercised the more vigorous type of policy entrepreneurship by contributing to the translation within the CAP of the conception holding that biotechnology is not compatible with consumer expectations in relation to organic food products. From around mid-1997, the EP Committee on Agriculture, the EP at large, the Commission and the Council all contributed to the carrying of this conception. There are weak indications that this conception was made available by people representing organised interests within organic farming (see below).

Policy entrepreneurship has also been exercised in relation to the establishment of a forum for communication. Agri-environmental policy – including organic farming – had already been envisaged as an area for closer cooperation between the DG for Agriculture and the DG for the Environment both by the Commissioner for Agriculture (Franz Fischler) and the Head of Unit for water protection, soil conservation and agriculture (Grant Lawrence) within the DG for the Environment (*Agra Europe* 5.5.1995). Along these lines, the first time the DG for the Environment and the DG for Agriculture came together to organise a conference (in 1999), it was on the issue of organic farming. Enabled by the Commission and EU financing, the DG for Agriculture and the DG for the Environment thus exercised policy entrepreneurship by both contributing towards – together with the Austrian Ministry for Agriculture, the Austrian Ministry for the Environment and the Government of Lower Austria – the establishment of a forum for the production of meaning on organic farming in the EU (Commission 1999).

Conflicts Over Boundaries – the Copenhagen Conference (2001)

Although the 1999 Conference does not immediately give rise to formal decisions within the Council, attention is drawn to the conference recommendations in the context of a Council meeting in October 2000 by the Danish delegation, who also called for a follow-up conference in the first half of 2001 (Council 2000). Moreover, while problems and conflicts relating to the boundaries of organic farming vis-à-vis other concerns were reiterated, still further concerns became added and the 'new' CAP was articulated as the forum where solutions should be pursued at the 1999 conference: these conceptual developments were confirmed and more widely accepted at the conference on organic farming in the EU in 2001.

The European conference on 'Organic Food and Farming – Towards Partnership and Action in Europe' was thus launched in May 2001 in Copenhagen, Denmark. The aim of the 2001 conference was to move towards the development of a European Action Plan on organic food and farming. At the conference Austria, Denmark, Estonia, Finland, Germany, Greece, Ireland, Lithuania, Norway, Sweden and the UK were represented by Ministers or Vice-ministers for Agriculture, and the Netherlands and Switzerland were represented by high level officials attached to the Ministry for Agriculture. The participants also included representatives from the DG for Agriculture, the DG for the Environment, COPA, IFOAM, EURO COOP (European Community of Consumer Cooperatives), the EEB, researchers, and State officials from across Europe (Danish Ministry for Food 2001). Compared with the 1999 conference, the range of participants was thus wider and included an additional number of those agents, more commonly considered to govern the CAP and favour the status quo, that is, Ministers of Agriculture and the DG for Agriculture and COPA. One outcome of the conference was the 'Copenhagen Declaration'. To be sure, neither of the 1999 Conference Summary Statements nor the Copenhagen Declaration are legally biding: yet they are declarations of intent, which creates certain expectations about future Community action in relation to organic farming. In particular, the Copenhagen Declaration may be considered to represent some degree of authority since it is signed by the participating Ministers and Vice-ministers for Agriculture, high level officials attached to the Ministry for Agriculture in Member States, and interest organisations.

It is broadly accepted that organic farming is a potential and partial solution to problems within the CAP. The Copenhagen Declaration, for instance, states that '[o]rganic farming is a highly relevant tool, which contains the potential to participate in solving simultaneously a range of problems related to food production, environment, animal welfare, and rural development' (The Copenhagen Declaration in Danish Ministry for Food 2001, p.6). From speeches at the conference it was qualified that organic farming constitutes a potential and partial solution to problems related to biodiversity, food safety and quality, consumer demands, less competitive rural areas and also the quality of life of farmers (Danish Minister for Food; Greek Viceminister of Agriculture; Austrian Minister for Agriculture; Vicepresident of COPA; Swedish Minister for Agriculture in Danish Ministry for Food 2001).

Like the 1999 Conference Summary Statement, the Copenhagen Declaration does not point to the sources of the current problems in agriculture but merely states that problems exist in relation to 'food production, environment, animal welfare, and rural development'. Of the speeches made at the conference, however, it appeared that there existed at least two and possibly three alternative conceptions of the sources of the current problems in European agriculture. One causal conception holds that the current problems are related to the working of the 'old' CAP.

The 'old' CAP aimed at increased productivity by means of the modernisation and industrialisation of the agricultural sector. A productivity focused CAP has, thus, led to the modernisation and industrialisation of agriculture which, in turn, has caused problems in relation to the environment, animal welfare and food quality (DG for Agriculture; German Minister for Consumer Protection; EEB in Danish Ministry for Food 2001). Another and alternative conception is that the current problems are caused by pressures from the market. Even though intensive agriculture is related to processes of modernisation, the functioning of the CAP is not the source of industrialised

European agriculture but instead it is caused by pressures from the market. That is, the source of problems in agriculture is conditioned by developments in a modern society but intensive agricultural production is a market-led development – not a product of the CAP (DG for Research in Danish Ministry for Food 2001).

A third conception – but not necessarily in conflict with either of the above – seems to hold that sources of some of the central problems of the CAP are to be found outside of the Community. More specifically, the sources of problems of food safety and quality are mainly expressed as being external to the Community. For instance, '[t]he anonymity of much of our food today has resulted in fears that it may have been produced with dangerous or unwanted additives, or in a way that some consumers would consider unethical. Surveillance of food production in the EU means that in order to ensure consumer satisfaction, food imports must also be monitored. The problems encountered when dealing with some of our major trading partners, regarding the use of hormones in beef production and with genetically modified organisms, are founded on the real concerns of European consumers and citizens. We must not employ double standards' (DG for Agriculture in Danish Ministry for Food 2001, p.42). Taken together, none of the three causal conceptions identifies the 'new' CAP as a source of current problems in agriculture. Moreover, like two years earlier, but now even more widespread, the CAP is identified as the forum, which should give further momentum to the organic farming sector.

Regardless whether it is the functioning of the 'old' CAP, pressures from the market or sources outside of the Community that are emphasised as the causes of current problems, it thus was broadly accepted that the CAP is the forum where solutions should be pursued (The Copenhagen Declaration; Danish Head of State; UK Junior Minister for Agriculture; DG for Agriculture; Vice-president of COPA in Danish Ministry for Food 2001). In other words, whereas there are conflicting conceptions as to the sources of current problems within the CAP, it is widely accepted that the changed or 'new' CAP is the forum where solutions to problems facing organic farming should be found.

Only rarely is this conception challenged. One such challenge was articulated by the President of IFOAM and holds that the existence of international, EU and national standards for organic production is problematic, however, in contrast to striking up a call for further unification of EU regulations, the importance of international dialogue is emphasised. That is, dialogue with parties outside of the EU and particularly in relation to trade negotiations in the WTO. Here the international forum (as opposed to the European one) is not seen merely a source of problems but, in fact, a possible forum where solutions should be pursued (President of IFOAM in Danish Ministry for Food 2001).

While organic farming has increasingly been articulated as faced with a series of problems throughout the current period, such problems are of central concern to the Copenhagen conference. Some of these problems have to do with fraud within the organic farming sector and a lack of consumer awareness about the environmental benefits of organic production. Most central are, however, problems related to the rules guiding organic production, which are linked to the distinctiveness of organic farming vis-à-vis other types of agricultural production. For instance, the existence of uneven standards for organic production within the EU, as well as between EU and third countries, is not only the source of trade difficulties but also formulated as a threat to consumer confidence in organic food products (Danish Minister for Food; President of IFOAM in Danish Ministry for Food 2001). Furthermore, growth in the organic farming sector at the expense of less strict rules for organic production is raised as a problematic potentially undermining the distinct identity of organic products. For instance, the Danish Minister for Food (and former Commissioner for the Environment) referred to a conception holding that 'growth can only be secured by compromises concerning definitions of the organic criteria such as: More additives allowed, more flexibility as regards animal welfare etc. etc. I understand the argument, and I fully agree that growth shouldn't be realized at the expense of clear and consequent principles. But I do not agree that this will happen' (Danish Minister for Food in Danish Ministry for Food 2001, p.30). Even though it is here rejected that the distinctiveness of organic farming necessarily needs to be compromised in the process of growth, the quote illustrates the existence of concerns related to the standards of organic food production.

The source of this problematic is clarified by a Commission guide to community rules on organic farming from 2001 elaborated by the DG for Agriculture. Of the document it appears that, 'conventional agriculture has been increasingly subject to strict environmental and animal welfare rules. This has meant the development of new approaches and methodologies, such as integrated agriculture'. Moreover, '[t]he organic farming sector now needs to see where it stands in relation to these new developments, and consider the production rules it applies with a view to maintaining a specific identity clearly distinguished from conventional agriculture' (Commission 2001, p.24). The success of organic farming thus depends on a clear definition of this mode of agricultural production and importantly such definitions involve a clear line of demarcation towards both conventional agriculture and integrated farming. A final conceived problem facing organic farming is related to the decreasing role of organic farmers' organisations in the elaboration of regulations addressing the organic farming sector. The President of IFOAM thus holds that:

[w]hile the organic sector had a substantial influence in the initial stages of the development of the regulation, this has gradually diminished. Drastically put: the right to define organic agriculture has been taken away from those who are practising it. The sectors' own initiatives, such as the IFOAM Accreditation Programme, have not been properly integrated. The new interest in the organic sector should not lead to more micro-management of the organic sector by national governments and EU regulations – we need less of that (President of IFOAM in Danish Ministry for Food 2001, p.34).

From within the organic farming movement, the organic farming sector is articulated, hence, as increasingly being excluded from the decision-making processes, which is also a problem voiced by representatives of the organic farming sector in the UK (see below).

Towards a European Action Plan for Organic Farming (Towards the End of the Period)

The EP Committee for the Environment had as early as in 1986 made a call for a European Action programme on 'biological farming' (European Parliament 1986, p.9). In the late 1990s, this call was reiterated in the context of the 1999 conference and, in particular, the Copenhagen Declaration following the 2001 conference states that organic farming 'should be developed further in Europe' through a European Action Plan. Additionally, in two years time, such a European Action Plan should '[c]over all aspects concerning the development of organic food and farming in Europe', involve 'all stakes-holders within Europe as a whole', and conclude in a 'consensus-orientated and market-based' strategy (The Copenhagen Declaration in Ministry for Food 2001, p.6).

At a meeting in July 2001, the Council agreed that the Commission should look into the support in favour of the elaboration of a European Action Plan (Council 2001). The DG for Agriculture was primarily responsible for the elaboration of the subsequent Commission working document, however, in order to search the ground for the possibilities for a European Action Plan on organic food and farming, an inter-service working group was established. Additionally, an expert group was set up consisting of representatives from a wide range of organisations. [18] Against this background, the Commission published a report in late 2002 entitled, 'Analysis of the possibility of a European action plan for organic food and farming' (Commission 2002). The Commission report endorsed a number of those problems facing organic farming articulated previously in the present period. These problems have to do with an information deficit both among consumers and within the production and processing chain, a lack of relevant statistical material, market transparency, the standards of organic produce from third countries, the working procedures between organic farming inspection bodies and government authorities, and

---

[18] This range of organisations included: IFOAM, the EEB, EURO COOP, European Organic Certifiers' Council (EOCC), COPA, COGECA, European Farmer Co-ordination (CPE), BEUC, Confederation of the Food and Drink Industries of the EU (CIAA), Comité Européen de Liaison des Commerces Agro-Alimentaires (CELCAA), Promoting Sustainable Rural Development in Central and Eastern Europe (Avalon), and a number of researchers and independent consultants (Commission 2001, p.4, Annex III).

research and training into wide number of aspects of the organic farming sector (Commission 2002, pp.25–25).

The report, however, also has at least two marked features. First, organic farming is suggested as a solution to a series of familiar problems within the CAP: however, the benefits of organic farming in this regard is argued on the basis of references to research addressing various aspects of organic food production. With references to recent research, it is thus argued that organic farming has beneficial effects on the protection of landscapes, wildlife, water quality, and diversity of flora and fauna. Moreover, organic farming is seen to work in favour of the protection of soil quality, animal welfare, the limitation of CO_2 emissions, and rural development by increasing economic activities and employment opportunities (Commission 2002, pp.27–28). Second, organic farming is in fact argued to be a solution to a number of problems within the CAP, which have risen out of recent changes in the objectives of this policy. That is, the Commission report claims that the CAP obtained new objectives in the course of its development and now, among other things, also aims to promote '[p]roduction methods which are environmentally friendly and able to supply quality products, [d]iversity in the forms of agriculture, product variety and the provision of public goods linked to rural development [and]…the provision of non-food (e.g. environmental and animal welfare related) services that the public expects from farmers' (Commission 2002, p.5).

The 2002 Commission report formed the basis of further consultations with Member State representatives and other stakeholders and the continuous work towards the elaboration of a European action plan for the organic sector was endorsed at a Council meeting in late 2002 where '[i]n general all delegations welcomed the Commission's working document' (Council 2002). Finally, after an 'online consultation', where all concerned were invited to voice their opinion on a number of issues regarding a European action plan (Commission 2003a) and a Commission Hearing (Commission 2004), a European plan containing 21 actions was issued by the Commission in mid-2004 (Commission 2004a) and, subsequently, 'broadly welcomed' by the Council (Council 2004).

Ideas Within Environmental Interest Organisations and IFOAM (mid-1990s)

Although environmental interest organisations such as Greenpeace and The World Wide Fund for Nature (WWF) are only marginally involved in carrying organic farming as a solution to agricultural problems in the EU, calls for support of the organic farming sector have also appeared from this quarter. For instance, in the first half of the current period, a joint report from the German Institute for Future Developments in Ecology and Greenpeace suggests that a complete conversion of German agriculture to organic farming is the most cost effective solution considering the costs of agriculture on the environment and animal health. The report is also intended for the EU Agriculture Ministers and was followed by a call for a transferral of CAP subsidies to support to organic farming over a period of eight years (*Agra Europe* 23.12.1992).

Along these lines, the WWF have proposed that the objectives of the CAP embedded in the Treaty of Rome are changed to match the 1990s wishes to promote environmentally friendly farming methods – rather than reflecting the problems of the 1950s. Hence, the rigidity of the CAP objectives drawn up more than 30 years earlier have been identified as a central source of lack of ability to resolve current problems in agriculture (*Agra Europe* 27.4.1990, 19.8.1994). In 1994, a report from the WWF and the Institute for European Environment Policy criticised the current CAP for not creating incentives to move towards less intensive farming – hereunder organic production. The newly appointed Commissioner for Agriculture (Franz Fischler) welcomed the report. The Commissioner for Agriculture, however, also held that the 1992 CAP reform 'pointed policy in the right direction' and the CAP should change by evolution rather than revolution (*Agra Europe* 24.3.1995). Whereas Greenpeace early on in the current period carried a link between organic farming and animal welfare, which is a conception that was institutionalised during the current period, a complete conversion of agricultural production to organic farming and constitutional change as suggested by the WWF in order to 'green' European agriculture was rejected.

Contrary to the environmental interest organisations, IFOAM seems to make ideas available for translation, or at least draw on some of the conceptions on the relationship between organic farming and the CAP, which was institutionalised within the CAP during the current period. IFOAM established an EU Working Group in 1990, which assisted the EP Committee on Agriculture in drawing up their response to an early Commission proposal for a Community regulation concerning the organic farming sector (European Parliament 1991a, p.76). Moreover, IFOAM participated in successive conferences (1996, 1999 and 2001) and a hearing (2004) on organic farming in the EU held during the current period, holds a membership of the Standing Committee on Organic Farming, and is recognised for its expertise in the elaboration of the European Action Plan on organic food and farming. In 2000, the IFOAM EU Working Group was replaced by an IFOAM EU Regional Group (IFOAM 2003).

Against this background, it appears that IFOAM possibly makes ideas available for translation, or at least draws on ideas, which are elevated to institutionalised concerns within the CAP. Accordingly, in the beginning of the current period IFOAM voiced the conception that the distinctiveness of organic farming is threatened by other types of agricultural production within the CAP. The problems facing organic farming are thus linked to the distinctiveness and very identity of organic farming which is thought to be under pressure from 'the financial and lobbying power of the chemical industry…and their allies like politicians [and] farmers associations' (IFOAM 1996, p.20). The rules setting up uniform standards for organic production are considered to set up the boundaries for organic farming but these rules must also be defended from pressure to allow the use of GMO's in organic farming and, in general, be continuously clarified against, for instance, integrated farming (IFOAM 1996, p.20). That is, early in the current period IFOAM gave voice to a number of problems, which was translated by the DG for Agriculture, carried by a wide number of agents and elevated to institutionalised concerns within the CAP by the end of the period.

Organic Farming and Policy Entrepreneurship in Member States (1993–2005)

During the current period organic farming attracted attention as an agricultural sector subject to considerable growth in several EU Member States including Denmark, the Netherlands, the UK, Spain, France, Germany, Sweden and Austria (the latter two attaining EU membership in 1995). There exists, however, some variations across these EU Member States on the nature of, and reasons given for, such growth. Likewise, there is some variation across EU Member States in the suggestions offered regarding the types of problems whose solutions are located in organic farming.

Denmark, the Netherlands and Spain (1993 onwards)

In Denmark growth in sales in organic food products is explained by a growing consumer interest in such products instigated by problems relating to groundwater pollution, the finding of agri-chemical residues in conventionally produced agricultural products and the campaigning for organic food products by major supermarket chains (*Agra Europe* 27.8.1993, 1.7.1994, 26.8.1994). Growth in the organic farming sector is also linked to the national support scheme for conversion to organic farming and expectations about potentially profitable export niche markets, especially in Germany and the UK (*Agra Europe* 8.1.1993, 13.1.1995). There also exists among farmers, however, some concern about the tenacious nature of the growth in markets for organic produce, both at home and, in particular, for export markets. A gradual adoption of stricter environmental requirements in conventional farming was voiced as a possible viable alternative to organic production (*Agra Europe* 13.1.1995). Significant in the production of discourse and the institutionalisation of problems and solutions in relation to organic farming is the Organic Food Council. The Organic Food Council was set up in 1987 by the first national regulation of organic farming but it gained considerably in importance during the current period, as it became the forum for the elaboration of two successive action plans directed at the development of organic farming in Denmark (Lynggaard 2001).

The first 'Action plan for the advancement of organic food production in Denmark' was thus published in 1995 (Structure Directorate 1995). The plan is elaborated and endorsed by a broad set of both conventional and organic farmers' organisations as well as by the Ministry for Agriculture. An 'Action Plan II – Developments in Organic Farming' followed this in 1999, which was elaborated and subsequently endorsed by an even broader spectrum of public and private agents including the main conventional farmers' umbrella organisation – the Danish Agricultural Council – and the Trade Council of the Labour Movement (Structure Directorate 1999). Whereas, the first Action Plan is a comprehensive account of problems and solutions regarding the further expansion of the organic market in Denmark, the concerns of the second Action Plan are the development of export markets, especially in Germany and the UK, and the link between organic production and consumer demands for food quality. The second action plan also calls upon the Minister for Food and the Organic Food Council to support the further development of organic farming in the context of the CAP and, in particular, in the context of the Agenda 2000 negotiations.

At the end of the current period, the Danish Ministry for Food and the Minister for Food may be identified as exercising the type of policy entrepreneurship that involves the establishment of a meeting place, which brings together agents and enables the production of collective meanings and conflicts over meaning. That is, the second European conference 'Organic Food and Farming – Towards Partnership and Action in Europe' held in May 2001, was not convened by the Commission but was instead organised by the Danish Ministry for Food and hosted by the Danish Minister for Food (Ritt Bjerregaard) – who was – until recently, Commissioner for the Environment (Danish Ministry for Food 2001).

In the Netherlands, the recent 1992 CAP reform is conceived of as an opportunity to motivate further expansion in the organic farming sector and organic farming is an important part of the development of sustainable agriculture as well as a potential and partial solution to problems of groundwater pollution caused by intensive agricultural production. These are the conceptions articulated in two reports calling for further support of the organic farming sector, that were presented to the Dutch Parliament during the first half of the current period. In 1992, the Minister of Agriculture (Piet Bukman) thus submitted a policy document to the Dutch Parliament proposing a series of measures in support of organic farming (*Agra Europe* 2.10.1992, 27.11.1992).

Four years later, in 1996, 26 consumers, environmental, nature and animal rights organisations made a common appeal to the Dutch Parliament and asked the government to take a number of initiatives to develop home markets and the promotion and marketing of organic food products in the Netherlands. In this report, organic farming was formulated as a non-polluting agricultural production method and it was suggested that a national target should be set up aiming for organic farming to constitute one-tenth of the total agricultural land by 2005. Retailers who were also called upon in the report to promote their advertisements of organic food products and to keep profits at the same level of conventionally produced food products – supported the expansion of organic production. At the same time, supermarkets were demanding a more consistent flow of high quality organic products and, in general, a more professional organic farming sector. Furthermore, the Minister of Agriculture (Piet Bukman) called for the organic farming sector to pay more attention to the development of awareness of organic farming among consumers. Finally, the main conventional Dutch farmers' organisation (LTO) was not as involved in, and supportive of, the call for the expansion of organic farming as, for example, their counterparts in Denmark and their response to similar national initiatives. The suggested national ambition of an organic farming sector constituting one-tenth of the total agricultural sector was conceived to be 'unrealistic' and consumer demand was identified as the critical factor, which sets out the limits for growth in the organic farming sector (*Agra Europe* 11.10.1996; Biologica 1996, Samenvatting).

From around 2000 towards the end of the period, organic farming in the Netherlands is, on the one hand, subject to government support and from various organised interests, on the other hand, developments in the organic farming sector are increasingly disappointing compared with government objectives. The consumer groups, environmental, nature and animal rights organisations, which made their first appeal to the Dutch government in 1996, published and made a second appeal in 2000. Here, the realisation of the target of an organic farming sector constituting one-tenth of the total Dutch agricultural sector was deferred to 2010 (Biologica 2000, p.2): an ambition that is taken on by the Minister for Agriculture (Laurens Jan Brinkhorst) and turned into government policy (*Agra Europe* 12.10.2001; Ministerie van Landbouw 2000, 2004). While the organic farming sector is reportedly subject to growth in terms of consumer demand, sales, number of organic farms and land area used for organic farming, farmers interest in converting to organic production is lacking behind and even decreasing towards the end of the period (*Agra Europe* 3.8.2001, 14.6.2002, 20.9.2002, 16.7.2004, 7.1.2005, 8.4.2005). The lack of interest among farmers is – by the main Dutch organic farming organisation: Platform Biologica – blamed on uncertainty about future subsidies, sales and – by the LTO – lack of access to price premiums during conversion to organic production (*Agra Europe* 3.8.2001, 7.12.2001, 20.9.2002).

In Spain, organic farming – as in the previous period – was expanding mainly in largely unpopulated areas with a widespread use of primitive farming methods. Increased consumer interest in environmentally-friendly food production, support for organic production adopted under the auspices of the CAP, and expanding export markets in other EU countries are seen as drivers of this growth (*Agra Europe* 17.6.1994, 11.3.2005).

The UK and France (1996 Onwards)

Albeit reported slightly later than in Denmark and the Netherlands, the organic farming sector is also attaining increased attention as an agricultural sector subject to growth in the UK, and the Soil Association predicts profitable opportunities for one-tenth of all UK farmers. The organic farming sector is expanding in terms of sales, market shares, consumer awareness and profitability for organic farmers, although it is considered problematic – among others by the MAFF – that a major part of the UK demand for organic food products is being met by imports (*Agra Europe* 28.6.1996, 4.4.1997, 19.9.1997, 15.1.1999, 26.1.2001, 23.11.2001, 30.11.2001, 11.1.2002, 15.2.2002, 17.9.2004; MAFF 1999, 1998, pp.1–3). During these times of growth, organic food production is conceived as being a response to public concern in relation to the quality and safety of food products brought about by the 'BSE crisis' (see above) and the use of biotechnology in conventional agriculture. Organic farming is also seen as a contribution to the environmental objectives of UK agricultural policy and in particular the pursuit of biodiversity (*Agra Europe* 26.9.1997, 9.4.1998, 11.10.2002, 8.10.2004; MAFF 1999).

Around 1997/1998 the UK Minister for Agriculture (Jack Cunningham) claimed a strong support for organic farming, and stated his intention – while the holder of the EU Presidency at the beginning of 1998 – to promote organic farming in the negotiations around the Agenda 2000 discussions (*Agra Europe* 9.4.1998, 26.9.1997). Moreover, the NFU was showing an increasing interest in the market for organic food products. For instance, the NFU set up a working group on organic farming issues and – with support from the MAFF and the Soil Association – it launched the publication *Focus on: Organic Farming*. In this regard, the Deputy President of the NFU (Tony Pexton) stated that '[a]lthough the NFU doesn't advocate any particular farming system we want to encourage producers to look to the marketplace and respond to what consumers want' (*Agra Europe* 4.4.1997).

At the same time, organic farming sector organisations are claiming to be disregarded in the policymaking process, for instance, during the adoption of a national Organic Aid Scheme early in the current period. British Organic Farmers (BOF) found the consultation procedure with the MAFF to be 'a charade' and that the recent CAP reform was bound to fail since agricultural support was still too high. The main claim by the BOF is that '[o]rganic farming offers a longer term solution to all the problems facing agriculture today, and should be seen by policy-makers as having a real role to play in the next stage of the reform process' (Director of the BOF Patrick Holden in *Agra Europe*, 20.8.1993). On the other hand, around 1997/1998, in the context of a Government review on support for organic farming in the UK, almost 350 organisations were invited to take part in a consultation process resulting in more than 100 replies putting forward their respective suggestions (MAFF

1999). Additionally, while organic farming is conceived of as offering a solution to a wide range of environmental problems and the maintenance of the countryside, MAFF also states that 'organic farming is one of the options available for securing environmental benefits and a "greener" EC Common Agricultural Policy' (MAFF 1999: Environmental Benefits). Together, this is both an example of the existence of conflicts over the potential of organic farming regarding the resolution of problems in agriculture, and about where to draw the boundaries that determine which agents should be included in, or excluded from, decision-making processes on issues related to organic farming.

Similarly to conceptions articulated from around the mid-1990s among others within the DG for Agriculture and among organised organic farming interests at the EU-level, the distinctiveness of organic farming and food products is also questioned in the UK. For instance, based on a report drawing on evidence given by a wide range of interested parties, the UK House of Commons Agriculture Committee questioned claims concerning food safety, animal welfare and the environmental benefits of organic farming, as well as claims about the superior quality of organic produce. It was emphasised that further research is needed into the matter in order to assure consumer confidence (House of Commons 2001: Reality v. perception; see also *Agra Europe* 23.2.2001, 11.10.2001). The UK House of Commons Agriculture Committee, on the one hand, has suggested that '[t]he [organic] industry may need to be less messianic and more marketing-oriented in its public presentation' (*Agra Europe* 26.1.2001). On the other hand, the Committee has expressed fears of the organic farming sector losing its 'traditional values' as it becomes commercialised and large companies and supermarkets gets involved. The report is subsequently endorsed by the UK government (*Agra Europe* 12.4.2001).

In late 2001, a Policy Commission on the Future of Farming and Food was set up with the aim to advice the government on how to promote an 'economic, environmental and social' sustainable agricultural sector and was at least partly conceived of as a responds to the 2001 Food and Mouth crisis. The work of the Committee resulted in the report on *Food and Farming: A sustainable future*, the elaboration of which involved a consultation process that generated responses from more than 1,000 individuals and organised interests. In the report, organic farming is, articulated, among other things, as a public good in terms of its contribution to biodiversity and a number of environmental benefits. Compared with the previous periods, it is still considered problematic that the UK marked for organic food products is being covered by imports. However, it also appears that this problem is even more pressing than previously. The predicted further growth in the number of consumers able and willing to pay for food considered to supply certain environmental benefits and – unlike previous periods – the conception that the organic food marked had reached 'a point where the term 'niche' surely no longer applies' is adding to the concern with the share of UK organic farmers of the market for organic food product (DEFRA 2002, p.71). The solution is a national action plan aiming to convert the current 30% market share of UK organic farming of the domestic market for organic produce to 70% by 2010 and the establishment of an Advisory Committee on Organic Standards, which replace the UKROFS and include an additional number of representatives of organic farming interests (DEFRA 2002a). By the end of the period, a government action plan addressing the further development of organic farming in the UK had not only been adopted and subsequently evaluated in terms of efficiency in England (DEFRA 2004), but a separate action plan had also seen the light of the day in Scotland (Scottish Executive 2003). In Wales a first action plan had already been adopted in 1999 (Welsh Organic Food Industry 1999), and a second plan followed in 2004 (Agri-food Partnership 2004).

Starting around the same time as in the UK, growth in consumer demands and sales for organic produce was observed in France (*Agra Europe* 22.9.1995, 16.1.1998). A survey shows that French consumers conceive of organic produce as quality food and choose these products for health and taste reasons. Organic production is, however, scattered and scarce and supplies in the main a very limited niche market (*Agra Europe* 22.9.1995). In early 1998, the French Ministry of Agriculture launched a five year plan for organic farming, which included support to farmers converting to organic farming, research and training, and the marketing of organic products, and also set up a national 'office for organic agriculture'. It is considered problematic that the French market for organic food products, to a significant extent, has been supplied by imports and, hence, the objective of the government plan is to put an end to the marginalisation of French agriculture vis-à-vis the booming French organic market (*Agra Europe* 16.1.1998). Towards the end of the period, even though consumption is still increasing, the French organic farming sector is stagnating in terms of its share of the area used for agricultural production and the total area used for organic production in France is even declining, which is a concern of not only be the National Organic Farming Association (FNAB) but also the Minister for Agriculture (Dominique Bussereau) (*Agra Europe* 15.4.2005).

Germany, Sweden and Austria (Towards the End of the Period)

Organic farming attracts attention in Germany as consumer demand for organic food products are increasing. A survey shows that consumers choose organic products mainly for health reasons and less so for nutritional, environmental protection and taste reasons (*Agra Europe* 18.8.1995). That is, at this point in time a link between organic farming and human health issues is expressed both among German and French consumers as well as within the EP (see above), yet it remains an alternative conception and has not been endorsed by the Commission and the Council.

Organic farming has been subject to public regulation in Germany since early on in the previous period and, from the mid-1990s, Germany is considered 'to be far ahead the rest of the Union' in terms of the consumption of organic products and the number of organic farmers, area farmed organically and the number of firms involved in the sector is subject to growth (*Agra Europe* 10.2.1995, 6.7.2001, 2.7.2004). A central and ongoing government concern, however, is related to the conceived fragmented nature of the organic farming sector in Germany reflected through the existence of a wide range organised organic farming interests, standards for organic production and labels for organic food products. The existence of more than 100 labels for organic food products was addressed – but not resolved – in 1997 when a new label was launched on the initiative of the Central Marketing Association for Agriculture (CMA) and the Association for Ecological Agriculture (AGÖL) with the support of the government: although an EU-wide label is still being backed by the German government. Another attempt to enhance consumer confidence and the distinctiveness of organic farming was made in late 2001/early 2002 by the introduction of a national label for organic food (*Agra Europe* 16.1.1997, 7.9.2001, 16.11.2001).

In the latter part of the current period, organic farming is by the German government not only announced as a central part of a general turn in German agricultural policy, but also articulated as a solution to some of the most pressing problems within the CAP including animal welfare, food safety and quality concerns. Against the background of the BSE crisis, which was particularly severe in the UK and which turned into a concern for the CAP from around 1996 onwards (see above), the German Minister for Agriculture (Karl-Heinz Funke) renounced the possibility of a similar situation appearing in Germany, which was declared a BSE-free area. After the first cases of BSE appeared in Germany in late 2000 (the first BSE infection was discovered on 24 November) the Minister for Agriculture and the Minister for Health (Andrea Fischer) suffered heavy criticism and both stepped down and resigned on 9 January 2001. Subsequently, the Ministry for Agriculture was changed to become the Ministry for Consumer Protection and Agriculture, and the Green Party's Renate Künast took up office (*Frankfurter Allgemeine* 27.02.01; *European Voice* 1.2.2001, 22.11.2001).

In a speech to the German Parliament, the new Minister for Consumer Protection claimed that '[t]he BSE scandal marks the end of the old type of agricultural policy' and went on to describe the new directions in German agricultural policy as aiming at 'quality instead of quantity' (Minister for Consumer Protection in Verbraucherministerium 2001). In this regard, the government wish to boost organic food consumption to the level of 20 per cent of the market within ten years (Minister for Consumer Protection in Verbraucherministerium 2001). The intended change in German agriculture away from industrial farming was backed by the Kansler (Gerhard Schröder) and by the main opposition party – the Christian Democrats (*Agra Europe*

12.1.2001, 9.2.2001a, 9.2.2001b, 6.4.2001a). The national farm umbrella organisation Deutscher Bauernverband (DBV) is more ambiguous and emphasises that German agricultural competitiveness must not be undermined (*Agra Europe* 12.1.2001, 12.1.2001a, 9.2.2001b). This problem was endorsed by a group of agri-economists (42 of them) in a letter to the government claiming that '[t]he unilateral promotion of organic production will jeopardise the competitivity of German agriculture and will leave it permanently dependent on state subsidies' and '[t]he "small and organic" route will prove to be a cul-de-sac' (*Agra Europe* 19.1.2001).

The expectations established regarding a turn in German agricultural policy was recognised, for instance, by the Commissioner for Agriculture (Franz Fischler) at an International Green Week in Berlin. The Commissioner for Agriculture claimed that '[w]hen it comes to the environment, improved quality and taking consumer interests into account in agricultural policy, Minister Renate Künast will not have to try and break down the doors in Brussels: they are already wide open' (Franz Fischler in *Agra Europe* 19.1.2001; also cited in *Financial Times* 19.1.2001, front page). Finally, the turn in German agricultural policy towards a focus on a considerable expansion of the organic farming sector through further regulation and support and the carrying of this aim into the CAP came into conflict with conceptions of the potentials of organic farming articulated in France as well as among organised interests in the UK. Hence, the French Agriculture Minister (Jean Glavany) stated that 'organic farming is growing in France, as it is everywhere else, but from there to reaching 20% is an objective which seems to me, I won't say sheer lunacy, but unattainable at any rate'. Organic farming is also 'not the universal miracle response to all the problems of European agriculture' (*Agra Europe* 16.2.2001). Likewise, rather than the further expansion and regulation of organic farming at the EU level being a solution to, for instance, animal welfare, food quality and safety issues, organised agricultural interests in the UK – the NFU – conceive of such as the source of problems related to the 'over-regulation' of farming (*Agra Europe* 23.11.2001). Together, these examples illustrate that organic farming at this particular point in time is, in fact, an issue of high political salience at the EU-level but it also acted as an illustration of the conflicts over the boundaries and potentials of organic farming vis-à-vis conventional agriculture.

Organic farming also received attention and growth in the two new EU Member States in this period – Sweden and Austria. In Sweden, an agri-environmental plan co-financed by the Commission was forwarded, a central objective of the plan being to promote the expansion of organic farming, with a growth take-up target of one-tenth of total agricultural land by 2000 (*Agra Europe* 1.7.1994a). Latter on the government target is revised and raised so to end at aiming to expand organic farming to make up one-fifth of the total agricultural land by 2005 (*Agra Europe* 12.11.2004, 11.2.2005; Jordbruksdepartementet 2004). Throughout the period, organic farming in Sweden seems in particular to be linked to issues of animal health and groundwater pollution and, in this regard, the expansion of organic farming is seen as a significant contribution to the reduction of the use of chemicals in agricultural production in Sweden (*Agra Europe* 5.5.1995, 1.2.2002, 4.4.2003). Towards the end of the period, doubts are expressed as to the prospects of meeting the government target of growth in the organic sector, the continuing persistence of consumer demands of organic produce, the amount of scientific evidence underpinning the environmental benefits of organic production and level of interest among farmers in converting to organic farming. Yet, by favouring support for environmentally friendly farming over conventional agriculture, the 2003 CAP reform is also among farmers forming the background for a renewed interest in conversion to organic production (*Agra Europe* 5.10.2001, 1.2.2002, 11.2.2005)

The Swedish EU presidency in the first half of 2001 may be seen as having exercised policy entrepreneurship by its contribution to the establishment of forums for communication as created in the context of the elaboration of the European Action Plan on organic farming. The Minister for Agriculture (Maragreta Wimberg) thus expressed it as an ambition of the Swedish presidency to support the further expansion of the organic farming sector in the EU. It was also during the Swedish presidency that the Council agreed to back the call of the 'Copenhagen Declaration' to elaborate a 'European Action for Organic Production and Organic Food' (*Agra Europe* 26.1.2001a, 20.4.2001; The Swedish Presidency 2003).

In Austria, the organic farming sector was expanding throughout the chief part of the period, making it one of the largest among the EU Member States relative to the percentage of land farmed by the conventional agricultural sector. Government and subsequent EU support of the sector and an expanding market for organic food products were identified as the main drivers of the growth (*Agra Europe* 5.5.1995, 16.5.1997a). By the end of the period, however, a lack of investment in marketing, combined with an increasing practice of subjecting organic products to conventional processing, means that organic food products are now conceived of as being less distinguishable from other agricultural products. Contributing to the blurring of the identity of organic products vis-à-vis other agricultural produce in Austria is that consumers, in general, tend to conceive agricultural production in Austria as being extensive and 'natural' (*Agra Europe* 9.2.2001).

Institutionalisations, Conflicts Over Boundaries and Policy Entrepreneurship (1993–2005)

The development in problems and solutions as they are formulated in and around the CAP in the period from 1993 to 2005 may be summed up as shown in Table 7.1.

Table 7.1: Problems and Solutions: The CAP and Organic Farming (1993–2005)

	CAP (end of previous period)	CAP (towards the end of current period)
Problems	• Surplus production, build up of intervention stocks and budget pressures • Unjust distribution of support among farmers and regions • Environmental depletion • Tension with trading partners	• Liberalised international markets • The management and complexity of the CAP • Lack of legitimacy of the CAP among the wider public • Environmental depletion, food safety, food quality, animal welfare
Causes	• Guaranteed prices and direct product subsidies • Intensive farming • Decline of value of US dollar and world market prices	• Price policy and international competition • Eastern enlargement • Inadequacy of the CAP to resolve current problems and need to obtain legitimacy among wider public • Intensive farming • Guaranteed prices and direct product subsidies
Type of Problem and Causality	• Mutation of cognitively instituted/normatively instituted problems • Multi-causal • Structural and thematic causes • Combination of causes inside and outside the CAP with emphasis on the former	• Analytical instituted problems: mono-causal; structural and thematic causes; combination of causes inside and outside the CAP • Normatively instituted problems: multi-causal; structural and thematic causes; causes found inside the CAP
Solutions	• Direct aid and support for extensive farming • Restrictive price policy • Rural development policy	• Restrictive price policy, decoupling of support from yields and direct aid • New balance between common management and increased decentralisation • Simplification of the CAP and a rural development policy
Solutions linked to Organic Farming	• Meet consumer demand for protection of environment (affordability) • Contribute to CAP objectives of environmental protection and maintenance of the countryside (acceptability)	• Meet consumer demand • Contribute to food safety and animal welfare, diversify and improve quality of food (affordability) • Contribute to CAP objective of rural development including improvement of less competitive rural areas, maintenance of the countryside, environmental protection (acceptability)

The period from 1993 to 2005 is characterised by a reorganisation of the cognitively instituted problems central to the CAP by the end of the previous period, into analytically instituted problems by the end of the present period. That is, experiences related to high price policies (and, in turn, surplus production, the build up of intervention stocks and budget pressures) and insufficient competitiveness are coupled with expectations of more liberalised international trade: this forms the basis for the articulation and institutionalisation of problems that relate to an inadequacy of European agriculture to profit from liberalised international agricultural trade. Moreover, even though the CAP still needs to address a number of normatively instituted problems by the end of the present period, this period is seen as being characterised by a reorganisation of such problems in comparison with the normatively instituted problems that were in place by the end of the previous period.

That is, the unjust distribution of support among farmers and regions and environmental depletion are still matters of concern. However, these concerns – together with a rising concern with the spread of animal diseases – are in essence seen, by the end of current period, as the sources of normatively instituted problems regarding a lack of legitimacy among the wider public. Finally, the present period is characterised by experiences within the CAP relating to previous Community enlargements being coupled with the forthcoming eastern enlargement, this forming the basis for the articulation and institutionalisation of problems regarding the management and complexity of the CAP. The period from 1993 to 2005 is also characterised by the formation of a policy field concerned with organic farming within the auspices of the CAP. The formation of a policy field concerned with organic farming during the current period is illustrated by the confirmation of organic farming as a solution to a number of familiar problems relating to consumer demands and rural development, including the protection of the environment, the maintenance of the countryside, and less competitive rural areas, as well as an additional number of problems. Such problems have to do with food safety, food quality and diversity, and animal welfare.

Particularly illustrative for the formation of a policy field concerned with organic farming during the period from 1993 to 2005 is, however, the articulation and institutionalisation of a number problems and conflicts over the boundaries of organic farming within the CAP. These problems and conflicts have to do with questions of what distinguishes organic farming from other concerns, what sorts of processes should guide decision-making on matters concerned with organic farming, and which agents should be included in, and excluded from, a policy field concerned about the regulation of the organic farming sector (see Table 7.2). Finally, and importantly, it is widely agreed that the problems facing organic farming should be resolved within the 'new' or changed CAP and, more specifically, it is considered that the development of organic farming in the EU should be approached through a common European Action Plan addressing the sector as a whole. To be sure, it is not, for instance, whether the use of biotechnology is included or excluded from the policy field concerned with organic farming, and nor is it whether the EP or organised organic faming interests have become more or less involved in the decision-making process, which is the central concern here. Rather, the argument is that a policy field concerned with organic farming is evolving during the current period insofar as the period is characterised by the articulation and institutionalisation of a number of problems and conflicts over which matters should be included and excluded, the role and legitimacy of various agents, and the types of procedures that should guide this field.

Table 7.2: Conflicts over Policy Field boundaries concerned with Organic Farming (1993–2005)

	Lines of conflict
Conflicts over the boundaries for what distinguishes organic farming from other concerns	• No biotechnology/biotechnology • Superior food quality/not superior food quality • Health objectives/no health objectives • Organic farming/conventional agriculture increasingly subject to strict environmental standards and integrated farming • The potential of organic farming is 10% of the total agricultural sector or more/organic farming remains a niche market
Conflicts over boundaries about the sorts of processes that should guide a field concerned with organic farming	• Consultation/co-decision • Increased involvement of organised organic farming interests in decision-making process/unchanged involvement of organised interests, including organised organic farming interests in the decision-making process • EU-wide Regulation/Member State and international rules
Conflicts over boundaries about which agents should be included and excluded from a field concerned with organic farming	• Increased inclusion of the EP/unchanged inclusion of the EP • Increased inclusion of organised organic farming interests/unchanged inclusion of organised organic farming interests

To be sure, the exercise of policy entrepreneurship, regarding the formation of a policy field concerned with organic farming within the CAP, is enabled by certain ideational and institutional changes within and around the CAP during previous periods. Additionally, it seems that the widespread conception of the existence of a crisis is conducive to the exercise of policy entrepreneurship during the current period. That is, from 1996 onwards, the BSE crisis formed the basis for the articulation and institutionalisation of a link between organic farming and food safety concerns. The Commission, the EP Committee on Agriculture, the EP Committee on the Environment, individual MEPs and the EP at large thus articulate and authorise organic farming as a potential and partial response to the BSE crisis. Within the Council, organic farming is linked to food safety concerns and, in the UK and later on in Germany, organic farming was linked in particular to the BSE crisis and consumer concerns with food safety issues.

Against this background, a series of agents may also be identified as translators, as establishing forums for communication, and as carriers of concepts and conceptions, which was institutionalised during the current period and which links organic farming to the CAP. In this regard, the concepts and conceptions institutionalised during the current period involve the establishment of organic farming as faced by a number of problems, and the establishment of the CAP as the forum where such problems should be resolved. Moreover, policy entrepreneurship has been exercised regarding the establishment of links between, on the one hand, organic farming and, on the other hand, problems relating to rural development (including: less competitive areas and the maintenance of the countryside), the production of quality food products, and the dissociation of organic farming from biotechnology.

Table 7.3: Policy Entrepreneurship and the Formation of a Policy Field concerned with Organic Farming within the CAP (1993–2005)

	Types of policy entrepreneurship		
Concepts and conceptions	Translators	Establishing a forum for communication	Carriers
Establish organic farming as faced by problems	• DG Agri.	• DG Agri. (report; conference 1996 – together with CEPFAR) • DG Agri., DG Env. and AT (Conference 1999) • DK (Conference 2001)	• EP Com. Agri.; EP Com. Env.; EP • Agri. Commissioner; Env. Commissioner; Commission
Establish the CAP as the forum where problems facing organic farming should be resolved	• DG Agri.	• DG Agri. (report; conference 1996 – together with CEPFAR) • DG Agri., DG Env. and AT (Conference 1999) • DK (Conference 2001)	• Agri. Commissioner; Env. Commissioner; Commission • AT; DK; EE; FI; DE; EL; IE; LT; NO; SE; UK; NL • COPA; IFOAM; EURO COOP; EEB
Linkage of organic farming to the fulfilment of CAP objective of rural development	• DG Agri.	• DG Agri. (report; conference 1996 – together with CEPFAR) • DG Agri., DG Env. and AT (Conference 1999) • DK (Conference 2001)	• Agri. Commissioner; Env. Commissioner; Commission • AT; DK; EE; FI; DE; EL; IE; LT; NO; SE; UK • COPA; IFOAM; EURO COOP; EEB
Linkage of organic farming to the production of quality food products	• DG Agri.	• DG Agri. (conference 1996 – together with CEPFAR) • DG Agri., DG Env. and AT (Conference 1999) • DK (Conference 2001)	• Agri. Commissioner; Env. Commissioner; Commission • EP Com. Env.; EP Com. Agri.; EP • DK; NL; UK; FR; DE
Dissociation of organic farming from biotechnology	• EP Com. Env.	• DG Agri., DG Env. and AT (Conference 1999)	• MEPs; EP Com. Agri.; EP • Agri. Commissioner; Env. Commissioner; Commission • Agriculture Council; UK

It should be noted that whereas there are sporadic indications that ideas are made available for translation among organised organic farming interests, in the main, concepts and conceptions translated within the CAP during the current period are either drawing on alternative ideas articulated in previous periods, or, it is unclear where such ideas are made available for translation. The exercise of policy entrepreneurship linking organic farming to the CAP in the period from 1993 to 2005 may thus be summed up as in Table 7.3. A wide range of agents may be identified as carriers of concepts and conceptions linking organic farming to the CAP. However, the more vigorous type of entrepreneurship, which contributed to processes of translation and the establishment of forums for communication and, hence, giving momentum to the institutionalisation of organic farming within the CAP, were exercised by the EP Committee on the Environment, and (much more particularly so) the DG for Agriculture.

8

Conclusions

This book started out with a concern about how institutional change may be captured and conceptualised within the CAP with a particular emphasis on the type of institutional change, which is ideational in nature. The following will start out by recapturing the general argument of this study. Hereafter follows a section on the discursive and institutional changes captured by the discursive institutional approach, and a section which provides a conclusion on the conditions for, and dynamics of, institutional change within the CAP. Finally, the discussion turns to the virtues and limitations of the discursive institutional approach pursued in an endeavour to capture and conceptualise institutional and ideational change.

Recapturing the Argument

It has been argued that the body of literature dealing with CAP reforms, on the one hand, provides great insights into the continuity of the CAP. The explanations of continuity offered are largely institutional and suggest that the CAP is resistant to change due to:

- the fragmented nature of the EU polity, with its dispersion of power and the existence of a wide number of veto players;
- the highly sectorised nature of the CAP;

- the privileged access of the farm lobby to political institutions both at the national and EU levels;
- the existence of a permanent 'free-rider' problem within the CAP;
- the constitutionalised nature of the CAP objectives (path-dependency);
- agricultural exceptionalism;
- the adoption of environmental concerns which are not common within the agricultural sector itself but, rather, used to legitimise continuing support of agricultural production;
- the close monitoring of the Commission by Member State representatives;
- the Agriculture Council acting as a strong defender of the status quo.

On the other hand, parts of the CAP reform literature have increasingly suggested that changes have indeed appeared within the CAP. Accordingly, attention has increasingly been given to change within the CAP caused by: (i) international free trade negotiations and the policy entrepreneurial role of the Commission; (ii) a diversification of the interests of the farm lobby and, likewise, a diversification of the general interest representation both at the European level and within Member States; (iii) a rising degree of autonomy of skilful individuals in favour of change; (iv) the influence of 'outsiders' on both international and EU negotiations on agricultural reforms; (v) the actual and the perceived 'negative externalities' of the CAP; and, (vi) the actual and self-imposed budget crisis. Taken together, the CAP reform literature offers a wide range of institutional explanations as to why the CAP has been highly resistant to change. Yet increasingly some studies have identified changes, which may be regarded as institutional and ideational, and while a number of explanations of change have been proposed, it has been argued that conceptualisations of change offered by the CAP reform literature seem to lag behind the empirical observations made about ideational change within the CAP.

Rather than brushing aside the explanations of change offered by the CAP reform literature, the most promising mechanisms for capturing and conceptualising change – the role of ideas, the working of policy entrepreneurs and crisis as a condition for change – were discussed in the context of the rational choice, historical and sociological institutionalisms. The rational choice, historical and sociological institutionalisms helped to flesh out the concepts and explanations of institutional change hinted at by the CAP reform literature but also added to the palette of conceptualisations of the dynamics and conditions of ideational and institutional change.

Regardless of the virtues of the rational choice, historical and sociological institutionalisms, it has been argued that such approaches to the study of ideational and institutional change have certain limitations. It was argued that:

(a) analytically predefined notions of the nature and direction of ideational change significantly limit the likelihood of capturing ideational change in a given political field;

(b) it was unclear how change in problem perceptions may be identified and described empirically and how such change may be related to institutional change;

(c) assuming ideas are coherent and well-defined entities, significantly limits the likelihood of identifying processes of diffusion or learning: at the same time, such processes are arguably emphasising continuity rather than change;

(d) in order to properly capture the nature and effects of a crisis situation on change, such situations should not be assumed to be non-cognitive factors and/or exogenous to the field under study;

(e) when attributing policy entrepreneurs with psychological predispositions or extraordinary skills, it tends to generate residual and ad hoc explanations of change.

To be sure, the current study has largely refrained from disregarding particular institutional approaches to the study of ideational and institutional change within the CAP against the background of empirical evidence. Rather, above all, the argument has been that rational choice, historical and sociological institutional approaches both in general and in the context of the CAP have their virtues while simultaneously creating certain limitations as to capturing and conceptualising institutional and ideational change within the CAP.

Against this background a discursive institutional approach to the study of institutional change has been proposed, which has a pronounced concern with the study of institutional change and, in particular, seeks to deal with ideational change. It was suggested that – rather than making assumptions about the nature and direction of change – the study of institutional change is, essentially, an empirical deed, and the dynamic and conditions of institutional change must be understood in a particular discursive and institutional context. It was suggested that change within the CAP may be captured by means of a discursive institutional analytical strategy (see Chapter 3). It was proposed that the dynamic of institutional change may be conceptualised along the lines of conflicts over meaning, processes of translation and policy entrepreneurship, and the existence of alternative discourse (seen as a necessary precondition for change) and ideational crisis as conducive to institutional change. Finally, it was suggested that the CAP constitute a critical case for the study of institutional change, and that the articulation and institutionalisation of organic farming within the CAP may reflect processes of change beyond relevance for organic farming and be potentially illustrative for the usefulness of the discursive institutional approach proposed to the study of institutional change.

Discursive and Institutional Changes Captured Within the CAP

Based on the discursive institutional approach and the analytical strategy proposed by the current study and the subsequent empirical analyses, the following conclusions on discursive and institutional change can be drawn.

The Pretext ...

To be sure, the discursive and institutional developments captured prior to 1978 do not constitute institutional changes within the CAP but rather forms the background for subsequent discursive and institutional changes within the CAP. In the late 1960s/early 1970s the CAP thus has two central concerns. First, it is problematic that the agricultural sector lagged behind society at large in terms of not taking a share in the welfare boosts enjoyed by others. This problem is normatively instituted in the sense that it is based on a coupling between, on the one hand, an actual but also a predicted further deepening of existing societal welfare disparities that disfavoured farmers and, on the other hand, an ideal conception holding that public policies ought to balance out such disparities. Second, an incipient agricultural surplus production is considered problematic. This problem is analytically instituted in the sense it is based on a coupling between, on the one hand, predicted developments in agricultural markets and, on the other hand, the consequences of previous price policies and certain structural characteristics of the agricultural sector.

The early 1970s saw the articulation of a world problematique in the context of the Club of Rome. This problem is normatively instituted in the sense that it is based on a coupling between, on the one hand, predicted exponential growth in populations, agricultural and industrial production, the consumption of non-renewable resources and pollution and, on the other hand, an ideal conception holding that there are limits to growth and emphasising the need to strive towards a 'global equilibrium'. Through a selective process of translation, this ideal conception forms the basis for the articulation and institutionalisation of agriculturally related environmental problems and solutions within the emerging EC environmental policy. That is, while a link is yet to be established between intensive agricultural production and environmental depletion within the CAP in the late 1960s/early 1970s, this link is translated within the emerging EC environmental policy where it is elevated into an institutionalised concern in the latter part of the 1970s.

The translation of the world problematique is selective in the sense that the ideas holding that public policies should be designed to strive towards a 'global equilibrium' and that technological progress is contributing to problems contained in the world problematique are not translated and institutionalised within the emerging EC environmental policy. Rather, it is upheld within the emerging EC environmental policy that solutions to environmental depletion should be pursued by means of technological innovation. Finally, even though marginal to the overall objectives of the emerging EC environmental policy and even if it is not institutionalised since further research is needed to establish the possible benefits of organic farming: a link had been articulated between organic farming (or biological farming which is the preferred terms at this point) and Community agricultural policy concerns within the emerging EC environmental policy by the late 1970s.

The CAP Before and After 1985

The CAP may both be described by certain discursive and institutional characteristics stretching beyond 1985 and by certain discursive and institutional characteristics that differ markedly before and after 1985. In other words, the CAP may both be described in terms of continuity and in terms of change around 1985. Accordingly, the CAP is both before and after 1985 faced by problems that relate to agricultural surplus production, budget pressures and the use of public funding in support of the already better-off farmers. These problems are cognitively instituted in the sense that they are based on two dissimilar couplings. First, problems that relate to agricultural surplus production, budget pressures and the use of public funding in support of the already better-off farmers, are generated by experiences based on a coupling between evaluations of developments predicted as early as in the late 1960s and actual developments in agriculture up to 1985. Second, the cognitively instituted problems are – both before and after 1985 – based on a coupling between, on the one hand, actual and further predicted agricultural surplus production, budget pressures and the disfavouring of certain farmers and agricultural regions in the Community and, on the other hand, an ideal holding that the CAP should strive towards equalising such imbalances.

At the same time, the period up to 1985 is also characterised by the institutionalisation of a number of normative problems related to rural exodus, structural diversification, intensive agriculture, and environmental depletion. That is, whereas such problems before 1985 were articulated as a concern in various contexts – for instance, within the emerging EC environmental policy, among people involved in alternative agriculture, and among groupings within the EP – problems relating to rural exodus, structural diversification, intensive agriculture and environmental depletion are, after 1985, institutionalised concerns within the CAP. These problems are normatively instituted in the sense that they are based on a coupling between, on the one hand, an actual but also a predicted further modernisation of agricultural production, increasingly severe budgetary constraints, and depressed markets, and, on the other hand, an ideal holding that the CAP should contribute to the protection of the environment.

Guaranteed prices and direct product subsidies are – both before and after 1985 – conceived of as the sources of some of the problems facing the CAP. However, unlike the time prior to 1985, it was acceptable in the post-1985 years to refer to the modernisation of agriculture and a choice of society in favour of a 'Green Europe' as contributing to problems requiring the attention of the CAP. Moreover, whereas technological progress before 1985 was conceived of as a 'fact of life' to which the CAP needed to adapt, after 1985 it was also acceptable within the CAP to refer to technological progress as a source of some of the problems that should be dealt with by the CAP. The introduction of a higher degree of co-responsibility with the aim to (re)establish an incentive structure that makes farmers more responsive to market developments and socio-structural policies are both before and after 1985 regarded as viable solutions to some of the problems central to the CAP. After 1985, however, it also became acceptable within the CAP to refer to solutions, which conceive agriculture as a part of the larger economy and as the protector of the environment. While it was still acceptable in the early 1980s to refer to the need to protect the CAP from international competition, this seems to have become a less viable solution, particularly in the latter part of the 1990s. That is, 1985 was not marked by change with regard to this specific institutional development and, it is unclear when, and how such changes may have come about.

With regard to organic farming, before 1985, it had been articulated as a potential and partial solution to certain problems in agriculture yet, after 1985, organic farming had come to resemble a linguistic field in which further articulations and institutionalisations may take place. That is, among people involved in alternative agriculture both at the community level and within certain Member States (France, the Netherlands and the UK), organic farming was – before 1985 – being articulated as a potential and partial solution to the overuse of energy and agri-chemicals in agriculture and as contributing towards an ecological balance needed in order to ensure sufficient food supplies in the long term. At the same time, it was regarded doubtful as to whether organic farming would address problems of surplus production and research into the scope and potential of organic farming was needed. After 1985, organic farming was also linked to problems within the CAP and – in particular – to concerns with food quality and consumer demands for organic produce. However, further research was needed into the link between organic farming and food quality.

The CAP Before and After 1992

The CAP may also be described both by certain discursive and institutional characteristics stretching beyond 1992 and by certain discursive and institutional characteristics that differ markedly before and after 1992. Accordingly, the CAP was – both before and after 1992 – faced with cognitively instituted problems of agricultural surplus production, budget pressures and the build up of intervention stocks. Although rural exodus and structural diversity seemed to be turned into concerns with the unjust distribution of support between farmers and regions, such normatively instituted problems are central to the CAP both before and after 1992. Likewise, the CAP – both before and after 1992 – was faced with normatively instituted problems related to environmental depletion. Thus, the problems facing the CAP before and after 1992 represent continuity, with the qualification that tension with international trading partners became an institutionalised concern of the CAP after 1992, as opposed to before.

Significant changes are found in the sources of, and solutions to, the problems the CAP needed to deal with before and after 1992. The modernisation of agriculture, economic recession, technological progress, community enlargement and a choice of society in favour of a 'Green Europe' are, before 1992, conceived as being the sources of problems central to the CAP. However, after 1992 problems within the CAP are rather explained by the system of guaranteed agricultural prices, direct product subsidies, a decline in the values of the US dollar and world market prices for agricultural produce. Importantly, problems that – before 1992 – are cognitively and normatively instituted have – after 1992 – mutated in the sense that intensive farming is institutionalised both as the source of problems related to surplus production, and environmental depletion.

The solutions preferred in the endeavour to resolve the problems central to the CAP also differ before and after 1992. Prior to 1992 it was considered that the CAP ought to deal with agriculture as part of a larger economy and as the protector of the environment. However, from 1992 such conceptions had been turned into specific solutions that related to direct aid to farmers and support of extensive farming, a restrictive price policy and a rural development policy. With regard to organic farming, whereas it made up a linguistic field before 1992, organic farming was, after 1992, institutionalised within the CAP as an object for community regulation, as fulfilling certain consumer demands and as contributing to the fulfilling of CAP objectives relating to the protection of the environment and the maintenance of the countryside.

The CAP Before and After the Late 1990s

As events become contemporaneous the estimations of the points in time, which may be marked by ideational and institutional change, are attached with increased uncertainty. In other words, the future may show that it would have been more appropriate to make a synchronic cut in time slightly earlier (or later) than done in the current context which, in turn, may give rise to refined conclusions as to changes in the discursive and institutional context. Still, the late 1990s may be described by certain discursive and institutional characteristics before and after this point in time.

Surplus production, budget pressures and the build up of intervention stocks is – both before and after the late 1990s – of concern within the CAP. However, whereas problems relating to surplus production, budget pressures and the build up of intervention stocks before the late 1990s are cognitively instituted, such concerns came to form part of analytically instituted problems after the late 1990s. After the late 1990s, the CAP is thus faced by problems relating to more liberalised international agricultural trade, and the uncertainty as to whether European agriculture will be able to profit from this development. This problem is analytically instituted in the sense it is based on a coupling between, on the one hand, predicted more liberalised international agricultural trade and potentially profitable international markets and, on the other hand, actual developments within the CAP, which has established an experience of a relationship between high price policies and insufficient competitiveness in the world market. The management and complexity of the CAP was also an institutionalised concern for the CAP to deal with after the late 1990s. This problem is analytically instituted and based on a coupling between the predicted enlargement of the EU to the east and a number of actual enlargements throughout the development of the CAP, which has established an experience of a relationship between Community enlargements, on the one hand, and administrative complexity and structural diversity, on the other hand.

Moreover, the unjust distribution of support among farmers and regions and environmental depletion are normatively instituted concerns of the CAP both before and after the late 1990s. However, the inadequacy of the CAP to resolve such concerns are, after the late 1990s, increasingly conceived of as contributing to the establishment of a negative image of the CAP in the broader public arena. The image problem of the CAP is normatively instituted and based on a coupling between, on the one hand, an actual but also predicted further inadequacy of the CAP to resolve current problems in agriculture and, on the other hand, an ideal holding that the CAP needs to obtain its legitimacy among the wider public. Finally, whereas food safety, food quality and animal welfare before the late 1990s are articulated as problems that the CAP ought to deal with, such problems are institutionalised concerns within the CAP after the late 1990s. High price polices and direct product subsidies, intensive farming, enlargements of the EU and unfortunate developments in world markets are identified as the sources of problems within the CAP both before and after the late 1990s. However, after the late 1990s, international competition, the inadequacy of the CAP to deal with problems in agriculture and the need for the CAP to obtain its legitimacy in the broader public arena are also regarded as the sources of problems central to the CAP. The solutions envisaged both before and after the late 1990s are a restrictive price policy, a decoupling of support from yields in favour of direct aid to farmers and a rural development policy. However, after the late 1990s it was also acceptable to refer to solutions, which take particular national or local conditions into consideration and which aim to simplify the management of the CAP.

Finally, organic farming is – both before and after the late 1990s – institutionalised as a solution to consumer demands for the protection of the environment and as contributing to the objectives of the CAP to protect the environment and maintain the countryside. However, after the late 1990s, it was also acceptable within the CAP to refer to organic farming as a solution to problems related to food safety, the diversity of food products, food quality and animal welfare. In general, whereas the CAP, prior to the late 1990s, was exclusively made up of a number of commodity regimes or market organisations, after the late 1990s a policy field concern with organic farming had been institutionalised within the auspices of the CAP.

A Discursive Institutional Conceptualisation of Institutional Change Within the CAP

From a discursive institutional perspective, the institutional changes outlined above may be understood as having been conditioned by the existence of alternative discourse and at times the existence of ideational crisis. Additionally, the dynamics giving momentum to institutional change at various points in time may be conceptualised along the lines of processes of translation, conflicts over meaning and policy entrepreneurship. The following will summarise the conditions and dynamics of institutional change within the CAP from 1980 to 2005 and briefly describe the condition and dynamic of institutional change within the emerging EC environmental policy in the 1970s.

It appears that the institutional changes outlined have all been conditioned by the existence of alternative discourse, that is, alternative articulations of ideas. In the early 1970s, institutional changes within the emerging EC environmental policy were, thus, conditioned by alternative ideas made available in the context of the Club of Rome. In the early 1980s, institutional changes related to the institutionalisation of environmental problems and solutions within the CAP was conditioned, partly, by alternative ideas made available in the context of the emerging EC environmental policy and, partly, by alternative ideas made available by people involved in alternative agriculture. In the late 1980s/early 1990s, institutional changes related most notably to the institutionalisation of intensive agriculture as a source of both cognitively and normative instituted problems and, hence, a mutation of such problems within the CAP was conditioned by alternative ideas made available, among others, by the EP Committee on the Environment and DG for the Environment. In the late 1990s, institutional changes relating to the institutionalisation of food safety, food quality and animal welfare was conditioned by alternative ideas, which – during the second half of the 1980s – had been articulated among others within the EP Committee on the Environment and the DG for the Environment. Finally, in the late 1990s, institutional changes related the institutionalisation of a policy field concerned with organic farming within the CAP was conditioned by alternative ideas made available – among others – by people involved in alternative agriculture and subsequently – after their translation within the CAP – among groupings within the EP.

Widespread conceptions of the existence of a crisis seem to have been conducive to certain institutional changes within the CAP. In the first half of the 1980s, institutional changes relating to the institutionalisation of environmental problems and solutions within the CAP appeared against the background of widespread concerns with the still present energy crisis and economic recession. Such concerns could be found within the DG for Agriculture, the Commission, the EP Committee on Regional Policy, the EP at large, the Council, and among people involved in alternative agriculture. In the second half of the 1990s, institutional changes relating to institutionalisation of food safety, food quality and animal welfare concerns as well as the institutionalisation of a policy field concerned with organic farming within the CAP appeared against the background of widespread concerns regarding the BSE crisis. Such concerns could be found among individual MEPs, within the EP Committee on Agriculture, the EP Committee on the Environment, the EP at large, the Commission, the UK, Germany and within the Council.

At various points in time, processes of translation, conflicts over meaning and policy entrepreneurship appear to have given momentum to institutional changes within the CAP. Processes of translation appear to have given momentum to change both in the 1970s within the emerging EC environmental policy and in the early 1980s within the CAP. In the 1970s, the translation of the world problematique within the emerging EC environmental policy gave momentum to the articulation and institutionalisation of the conception that agricultural production is the source of certain environmental problems. The process of translation is selective in the sense that the solutions contained in the conception of a global equilibrium did not become adopted within emerging EC environmental policy. Likewise, the process of translation is selective in the sense that the conception that technological progress is not containing viable long-term solutions, did not, in fact, displace the conception that solutions to problems of environmental depletion should be pursued by means of technological innovations within the emerging EC environmental policy.

Moreover, in the early 1980s, a process of translation gave momentum to the articulation and institutionalisation of environmental problems and solutions within the CAP. The process of translation is selective in the sense that the articulation and institutionalisation of the conception that the modernisation of agriculture is a source of environmental depletion is selected from the emerging EC environmental policy. Likewise, the process of translation is selective in the sense that the conception that technological progress is the source rather than the solution to current environmental problems is selected from among people involved in alternative agriculture (whom, in turn, draw on the conception of the

existence of a world problemtique and a global equilibrium). The translation of
the conception that technological progress is the source rather than the solution
to current environmental problems within the CAP may also seen as
contributing to the displacement of the conception that technology is a 'fact of
life' to which the CAP must adapt. However, it should be noted that
technological progress is, even in the late 1960s, being articulated as the source
of certain problems within the CAP. It is, hence, perhaps more accurate to claim
that the process of translation in the early 1980s contributed to the enforcement
of an already existing conception within the CAP holding that technological
progress is a source of problems in agriculture. Finally, the process of
translation within the CAP in the early 1980s does not dislocate the cognitively
instituted problems related to surplus production and budget pressures but,
rather, the normatively instituted problems relating to environmental depletion
was institutionalised alongside existing concerns within the CAP.

Conflicts over meaning gave momentum to change in the second half of the
1980s and during the period from 1993 to 2005. The institutionalisation of
normatively instituted problems alongside cognitively instituted problems within
the CAP by 1985 in essence provided the basis for a number of subsequent
conflicts over meaning. Such conflicts over meaning evolve around the sources
of the normatively instituted problems, their solutions, and the priority to be
given to cognitively and normatively instituted problems. In the second half of
the 1980s, normatively instituted problems related to, for instance,
environmental depletion and rural exodus are conceived by some as caused by
intensive agriculture, which in turn is seen to be caused by technological
progress, urbanisation, and industrial developmentst. Others regard the
operational principles of the CAP – including the pursuit of market unity, a
preference for Community production and financial solidarity among Member
States – as the basic sources of normatively instituted problems within the CAP.
Again others, question the link between modern agricultural production methods
and environmental depletion as well as the scope of possible environmental
depletion in this regard. Some question the availability of Community solutions
to the possible problems that relate to environmental depletion and the beneficial
effects of a restrictive price policy. However, to varying degrees, the solutions
most often envisaged to resolve problems within the CAP – in the second half of
the 1980s – are seen as a restrictive price policy, direct aid to farmers and a CAP
in support of extensive farming and environmental protection.

In summary, in the second half of the 1980s there existed a high degree of fit
as to the problems that it was considered the CAP ought to deal with, although
the priority to be given to normatively and cognitively instituted problems
varied. Moreover, some degree of fit can also be seen to exist as to the solutions
to apply, although the priority to be given to the support of an extensification of
agricultural production and environmental concerns varied. On the other hand,
however, conflicts appear over the sources of current problems in the second
part of the 1980s. Such conflicts seemed to be tapering away in the early 1990s

as the cognitive and normative instituted problems mutated. The cognitively and normatively instituted problems mutate in the sense that intensive farming is articulated and institutionalised as the source of both problems relating to surplus production and environmental depletion in the early 1990s.

The conflicts over meaning, identified by the current study during the period from 1993 to 2005, also appear to have given momentum to the formation of a policy field concerned with organic farming within the CAP. First, conflicts appear over the boundaries as to what distinguishes organic farming from other concerns within the CAP, including: whether or not biotechnology is compatible with organic production, whether or not organic produce represents a superior food quality and, whether or not the regulation of the organic farming sector has health objectives. Conflicts also appear over the degree to which organic farming is distinguishable from integrated farming and conventional agricultural production, which is subject to increasingly strict environmental standards, as well as appearing over the potential scope of organic farming vis-à-vis the total agricultural sector. Second, conflicts appear over the boundaries for the type of processes that should guide a policy field concerned with organic farming. Areas of potential disagreement that invoke conflict include whether or not formal decision-making should proceed according to the consultation or co-decision procedure, whether or not organised organic farming interests should be increasingly involved in decision-making, and over determining the type of relationship between EU, national and international regulation of organic farming. Third, conflicts appear over the boundaries as to which agents should be included and excluded from a policy field concern with organic farming and, in this regard, the prominence of the EP and organised organic farming interests within this field. Finally, although conflicts appear along the lines of the above, there also exists a high degree of fit on the conception that solutions to problems and conflicts in organic farming should be pursued within the new or changed CAP.

Policy entrepreneurship appears to have given momentum to change in concepts and conceptions linking organic farming to the CAP at various points in time since the late 1970s. First, in the late 1970s/early 1980s, those involved in alternative agriculture, individual MEPs and, in particular, the EP Committee on Regional Policy was translators of a number of concepts and conceptions, which are not institutionalised within the CAP, but which gives momentum to the formation of what resembles a linguistic field linking organic farming to the CAP. These conceptions include organic farming as a potential object for community regulation and the articulation of links between, on the one hand, organic farming and, on the other hand, problems within the CAP, food quality issues, and consumer demands. Second, an individual MEP, the EEB and, in particular, the EP Committee on Regional Policy enable the articulation of links between organic farming and the CAP through their contributions to the establishment of a number of forums for communication. Finally, although various carriers draw on the conceptions outlined above none of these conceptions are authorised and, hence, institutionalised. Most notably, the Commission is a non-carrier of the conception that organic farming should be an object for Community regulation and both the Commission and the Agriculture Council called for further research into the possible benefits of organic farming.

When looking at the late 1980s/early 1990s, the EP Committee on Agriculture and, particularly, individual MEPs, and the EP Committee on the Environment was translators of a number of concepts and conceptions, which gives momentum to the institutionalisation of organic farming within the CAP. These conceptions include organic farming as a sector for Community regulation, as a solution to problems within the CAP, as in demand among consumers, and as contributing to the CAP objectives of protecting the environment and maintaining the countryside. Individual MEPs and, in particular, the EP Committee on the Environment, enable the production of meaning – and conflicts over meaning on links between organic farming and the CAP – through their contributions to the establishment of a number of forums for communication. Finally, as carriers the EP, the EP Committee on Agriculture, the Commission, the DG for Agriculture, the DG for the Environment, the Council for Agriculture and a number of Member States have all contributed to the institutionalisation of the conceptions linking organic farming to the CAP as outlined above.

From around the mid-1990s, the EP Committee on the Environment, and (much more particularly so) the DG for Agriculture, was translators of a number of concepts and conceptions, which gives momentum to the formation of a policy field concerned with organic farming within the CAP. The EP Committee on the Environment contributed to the translation of the conception that organic farming is not compatible with biotechnology. More notably, however, the DG for Agriculture contributed to the translation of the conceptions that organic farming is faced by a number of problems, that the CAP is the forum where such problems should be resolved, that organic farming contributes to the fulfilment of CAP objectives related to rural development, and that organic farming is a high quality food production method. Following on, Austria, Denmark, the DG for the Environment and, in particular, the DG for Agriculture enabled the production of meaning on, and conflicts over the boundaries of, a policy field concerned with organic farming within the CAP through their contributions to the establishment of a number of forums for communication. Finally, as carriers the EP, the EP Committee on Agriculture, the EP Committee on the Environment, MEPs, the Commission, the Agriculture Commissioner, the Environment Commissioner, a wide range of Member States, the Agriculture Council, IFOAM, COPA and other organised interests to varying degrees, all contributed to the formation of a policy field concerned with organic farming within the auspices of the CAP.

Reflections on the Virtues and Limitations of the Discursive Institutional Approach

Importantly, the discursive institutional analytical strategy pursued in the study of ideational and institutional change is not thought of as necessarily a competing but, rather, a potentially complementary analytical framework vis-à-vis rational choice, historical and sociological institutional approaches to the study of institutional change – and continuity for that matter. In that sense, it is acknowledged that several institutional logics may be at stake in any given institutional context. Although, theoretical integration may not be possible – or even desirable – the pursuit of the study of change within the CAP – and elsewhere – through various institutional frameworks may thus generate unlike but complementary aspects of processes of institutional change and, hence, contribute to a more comprehensive and in-depth understanding of the conditions for, and dynamics of, institutional change. It is in this vein the conceptualisations of the discursive institutional analytical strategy have been elaborated. It is also in this vein the following will seek to point to a number of challenges to, and possible contributions of, the discursive institutional perspective on the study of institutional change and, in particular, on the study of the type of institutional change, which is ideational in nature.

1. It appears that the proposed discursive institutional approach has been able to capture and conceptualise certain institutional and ideational changes within the CAP, which is otherwise commonly considered to have been very resistant to change. This has essentially been enabled by a reconsideration of the concept of institution and a reconsideration of the concept of change. Both conceptual reconsiderations arise out of the point of departure of the discursive institutional approach in a logical sequence between ideas, discourse, and institutions. Accordingly, institutions are thought of as authorised and sanctioned discourse, which in turn is thought of as articulated ideas that have been turned into rules-based systems of concepts and conceptions. Change is thus considered to appear as ideas are turned into discourse and as discourse is turned into institutions. The process of ideas being turned into discourse is one of articulation – the outcome of which is discursive change – and the process of discourse being turned into institutions is one of institutionalisation – the outcome of which is institutional change. However, if we reconsider what may serve as an institution, do such institutions still have institutional characteristics? Likewise, if we reconsider what may qualify as change is it still change, which is worthy of its name?

I believe the answer to both questions is yes. Regardless of the preferred institutional optic, the most basic characteristic of institutions is that institutions have to do with rules. It is also agreed that – whichever type of rules are in focus – institutions in one way or another contribute to the shaping of policy outcomes. It may be emphasised how the interaction between individuals and institutions shapes policy outcomes, how institutions structure political activity, shape behaviour and preferences and have a direct impact on policy outcomes, or how agents interpret and give meaning to the world through institutions. The discursive institutional approach pursued in the current study is concerned with the rules that govern discourses, and institutions are seen as the devices through which agents interpret and give meaning to the world. Along these lines, institutions are considered to establish expectations about what is viable but also non-viable political activity in a particular context. By constituting a set of authorised and sanctioned discursive rules, institutions delimit, for instance, acceptable and valid statements from those which are unacceptable and invalid, as well as guide the formulation of relevant problems and their solutions in a particular context. While not disregarding the importance of formal decisions, for instance, made in the Agriculture Council for the particularities of the outcome of initiatives taken within the CAP, the discursive institutional approach pursued in the current study points up the significance of the formation of political problems and solutions, which tend to take place long before any formal decisions are made. The articulation and institutionalisation of certain problems, their causes and solutions rather than others, thus set out a space of possibility, which allows for certain initiatives and solutions to be suggested in the first place, while disregarding others and, hence, in a significant way, guides political activity and delimits potential policy outcomes.

2. Rather than making a priori assumptions about the nature of change within a given field or approaching the study of institutional change by means of analytically predefined ideas against which change is measured, the articulation and institutionalisation of ideas and, hence, discursive and institutional change is essentially an empirical question for the discursive institutional approach. Through an analytical inductive methodology, as proposed by the discursive institutional approach to the study of institutional change, it has thus been possible to capture ideational and institutional changes within the CAP, which other institutional optics are often unable to detect, and where empirical studies of the CAP can only point to a suspicion of change. For instance, it has else where been observed that the CAP has been subject to a 'thin and unstable' institutionalisation of environmental concerns. Yet a 'thin and unstable' institutionalisation is commonly registered not as a change and, hence, not sought explained as a change – rather it is registered as an example of the lack of change. In contrast, the discursive institutional approach would register and seek to explain even a 'thin and unstable' institutionalisation of environmental concerns within the CAP. And – possibly – this is the most typical type of change that may be observed in policy fields otherwise characterised by a high degree of continuity.

3. It has not been the ambition of the current study to establish causal relationships between, on the one hand, the nature of a discursive and institutional context and, on the other hand, the particularities of policy outcomes or courses of action. However, it could be argued that further theorisation over such relationships is indeed needed. The articulation and

institutionalisation of a number of links between organic farming and the CAP over time could be seen as made possible by changes in the problems with which the CAP has been expected to deal. That is, as ideational change appears, new types of policy instruments become relevant. The empirical analysis suggests that, for instance, as environmental depletion and later food quality issues had become institutionalised concerns within the CAP, the way had been paved for the institutionalisation of, for instance, organic farming as a solution to such problems. At the same time, however, organic farming as a solution to certain problems – and no doubt other issues – seems to have contributed to the articulation of problems within agriculture, which are later elevated to institutionalised concerns within the CAP. In other words, the empirical analysis suggests that the relationship between the discursive and institutional context and particular courses of action is not a simple one based on cause and effect but a complex one of interrelating causes and effects, which needs further theoretical reflections than has been presented in the current context.

4. From a discursive institutional perspective it has been proposed that the existence of alternative discourse is a necessary condition for institutional change since it is only through disputes over the articulation of ideas that the existing institutional context is contested. The most important implication of this notion is probably the question of how we may be able to distinguish one discourse from another. That is, how is it possible to pinpoint a boundary, which delimits one discourse from another? This question has not been fully dealt with in the preceding analysis. For instance, the world problematique, as articulated in the context of the Club of Rome or the emerging EC environmental policy in the 1970s, may not constitute separate meaning systems in a strict sense. However, in the 1970s, the ideas voiced in the context of the Club of Rome or the emerging EC environmental policy does represent alternatives to those articulated and institutionalised within the CAP. Essentially, rather than referring to full-blown alternative meaning systems, the current study has pointed to alternative articulations of ideas either in other contexts/policy fields than the CAP or to alternative non-institutionalised ideas articulated within the CAP.

5. It has been proposed that the existence of widespread conceptions of crisis is conducive to institutional change and it appeared from the empirical analysis that such conceptions did, in fact, form the basis for institutional change both in first half of the 1980s and in the late 1990s. However, empirically it also appears that it may at times be difficult to pinpoint the boundary between a problem and an ideational crisis. For instance, in the late 1960s, problems relating to agricultural surplus production were analytically instituted in the sense that such problems were based on a coupling between, on the one hand, predicted developments in the agricultural markets and, on the other hand, insights obtained regarding the consequences of previous price policies and certain structural characteristics of the agricultural sector. However, in the 1980s, problems of agricultural surplus production were cognitively instituted and had

moved up on the CAP agenda, and concerns with such problems had become more widespread and, arguably, resembled an ideational crisis. Although an ideational crisis is more voluminous in the sense that it involves a high degree of concord of its existence among the involved agents, empirically, there seems to be a gradual transition from a problem 'merely' being a problem to a problem having been turned into a crisis.

Moreover, and more importantly, the conceptualisation of crisis as an ideational condition for institutional change differs from the other conceptualisations proposed here by not being a concept, which essentially arises out of the particularities of discursive rules. Opposed to processes of translation, conflicts over meaning and policy entrepreneurship, the concept of ideational crisis has not been expounded in terms of its discursive characteristics. Arguably, a further refinement of the concept of ideational crisis in terms of its discursive characteristics is needed not only to improve the conceptual consistency of the discursive institutional approach proposed but also in order to clarify how an ideational crisis differs from, for instance, problems. Here it shall only be indicated that the refinement of the discursive institutional concept of ideational crisis may possibly find inspiration in the historical institutional proposition that institutional crisis might arise out of the accumulation of contradictions. However, whereas historical institutionalists may focus on political and economic contradictions, a discursive institutional approach would focus on discursive contradictions. For instance, the empirical analysis showed that the institutionalisation of normative problems in the mid-1980s alongside already existing cognitive instituted problems, gave rise, in the first instance, to a number of conflicts over meaning and, subsequently, a mutation in the sense that these problems were conceived of as being caused by the same phenomenon (intensive farming) by the early 1990s. However, it is at least conceivable that in situations where a mutation between non-uniform articulations of ideas is not observed, or is perhaps not even an option, severe discursive contradictions within a given policy field could give rise to an ideational crisis.

6. How may we conceptualise the internalisation of externally triggered ideational change? From a sociological institutional perspective, it has been suggested that such processes may be studied as the inflow and diffusion of new ideas. The conceptualisation of diffusion processes is no doubt a helpful description of how ideas are carried from one context to another in certain instances. However, sometimes the carrying of ideas from one context to another is less orderly and may perhaps rather be described as a process of translation. Particularly in the case of the CAP, it may be argued that: if ideas embedded in this field enjoy a high degree of institutionalisation then it may be expected that the introduction of new ideas does not happen easily. Thus, the concept of translation enable us, for instance, to capture the partial and selective institutionalisation of ideas, where certain ideas already embedded in the field up for study may be upheld fully, partially or possibly mutate with the new ideas being introduced. This line of thinking, which seeks to avoid externalising

explanations of change, also gave rise to an empirical analysis that does not claim a priori, for instance, that Community enlargements or relationships with international trading partners are exogenous to the CAP.

Although the involved agents may hold, for instance, that the sources of certain problems necessitating CAP attention stem from outside of the CAP, the actual conceptions very much form part of the CAP as a policy field as understood through the discursive institutional optic. Translations are thus seen as endogenous and ongoing processes conditioned by contacts with other policy fields or social contexts. For instance, to varying degrees, Community enlargements have given rise to various problems within the CAP for the greater part of the time since 1980. Likewise, relationships with international trading partners have been institutionalised concerns of the CAP since the early 1990s. That is, alongside, for instance, issues related to surplus production and environmental depletion, Community enlargements and relationships with international trading partners form part of the common concerns among a set of identifiable agents and are dealt with along the lines of commonly recognised processes within the CAP (cf. definition of a policy field in Chapter 3). To be sure, Community enlargements and relationships with international agricultural trading partners are not primarily seen as forming part of the CAP due to the definition of a policy field drawn upon in the current study. Rather, it appeared in the empirical analysis that conceptions relating both to Community enlargements and relationships with international agricultural trading partners have, in fact, been guiding for some of the problems and solutions central to CAP during certain periods of the existence of this policy field. Essentially, the concept of translation and the empirical research have suggested that perhaps – when it comes to the study of ideas – the CAP is not as sectorised a policy field as it is commonly considered to be.

7. From within the historical institutionalism it has been suggested, for instance, that policy-oriented learning in a policy sub-system may give rise to ideational change. Learning processes are here seen as, among other things, the product of competing advocacy coalitions and conflicts between beliefs systems. This approach to the study of endogenously generated ideational change is no doubt rewarding in cases where competing and institutionalised advocacy coalitions and belief systems already exist. However, it appears less appropriate in cases – such as the institutionalisation of organic farming within the CAP – which is characterised by – at least at first – less well-established coalitions and non-institutionalised belief systems. The discursive institutional approach thus suggest that in such situations it may be helpful to zoom-in on conflicts over meaning rather than on institutionalised beliefs and coalitions.

The proposed conceptualisation of a dynamic of institutional change relating to conflicts over meaning, appeared particularly useful in capturing conflicts over meaning in the second half of the 1980s. The conceptualisation of conflicts over meaning also appeared helpful in capturing conflicts on the issues to be included and excluded, on the agents to be included and excluded, and on the processes that should guide the policy field concerned with organic farming as it was evolving during the period from 1993 to 2005. Arguably, however, the quality of the conflicts over meaning identified throughout the period from 1993 to 2005 differs from that identified in the second half of the 1980s. On the one hand, both conflicts over meaning identified in the second half of the 1980s and during the period from 1993 to 2005, involve conflicts over the articulation of problems, their sources and solutions. On the other hand, however, and contrary to the conflicts over meaning identified in the second half of the 1980s, the conflicts identified during the period from 1993 to 2005 also involve conflicts over the inclusion and exclusion of agents as well as over the processes that should guide the policy field at hand. Arguably, the conflicts over meaning identified in the second half of the 1980s have consequences in terms of the actors, who may or not be accepted within the CAP, and for the processes guiding the CAP. It could also be argued that further empirical investigation may possibly reveal conflicts over the type of agents to be included and excluded, as well as the processes that guided the CAP in the second half of the 1980s. However, it seems that the quality of the conflicts over meaning, and conflicts over borders, somehow differs and further theorisation is need on this matter. Additionally, it should be noted that conflict over meaning is a concept that seeks to capture an endogenous dynamic of change in the meaning system or discourse through which agents are assumed to interpret and give meaning to the world. In that sense, the formation of preferences is also considered endogenous to such meaning systems. However, the discursive institutional approach to the study of institutional change, fall short of specifying the exact nature of the interrelationship between, on the one hand, the discursive and institutional order and, on the other hand, the formation of preferences. It has been suggested elsewhere that the discursive and institutional order 'shape' preferences, yet additional theorisation is needed in this regard.

8. Against the background of the CAP reform literature, the CAP has been assumed to make up a policy field, which may be distinguished from other fields in terms of a series of disputes around a common concern among a set of agents that operate through commonly recognised processes. This field definition has led to, for instance, considering successive enlargements and international trade issues as endogenous to the CAP insofar as disputes over such issues involve agents and processes familiar to the CAP. The conclusions made, should be seen in the light of the definition of a policy field drawn upon and are not readily comparable to the most common distinctions made in the CAP reform literature about which issues, agents and processes should be considered endogenous and exogenous in terms of the CAP. However, the field definition used has not been adopted in order to increase confusion and has not been chosen randomly. The general point is that discursive and institutional change – when thought of along the lines of the discursive institutional approach proposed – appears through processes of articulation and institutionalisation.

For instance, Community enlargements and international trade negotiations are institutionalised concerns to the extent that they have become so, through processes of articulation and institutionalisation within the CAP. This line of thinking, on the one hand, differ the most from explanations of institutional change as a product of abrupt external factors or shocks. On the other hand, although it is only hinted at in the current context, the line of thinking arising out of the discursive institutional definition of a policy field may differ less from theoretical frameworks that explain change with reference to developments and links between various levels of negotiation (cf., for instance, Patterson 1997; Coleman and Tangermann 1999).

In relation to the concerns raised about the definition of a policy field, it should be noted that a field description is a comprehensive task and, arguably, particularly so when it comes to the CAP. Whereas the current study has offered a description of an emerging field concerned with organic farming within the CAP, it has only sporadically pointed to certain characteristics of the CAP field, which have otherwise been assumed to exist. The latter is particularly important in terms of the conclusions drawn in relation to processes of translation. The conceptualisation of translations holds that the progress of such processes in one social context or policy field is conditional on contacts with other social contexts or policy fields. Consequently, in order to identify the translation of concepts and conceptions in a given field, then the first consideration has to be that such conceptions must not – before translation – form part of the discursive and institutional order of this field. It should also be possible to identity the field, which has made the concepts and conceptions available for translation.

The empirical analysis of processes of translation in the 1970s and the early 1980s has thus identified both concepts and conceptions, which in the first place were not part of the discursive and institutional order within which translations were taking place, and also identified the contexts, which made certain concepts and conceptions available for translation. Moreover, it has been indicated that certain concepts and conceptions translated within the CAP in the period after 1993 may have been made available by people involved in alternative farming. However, this conclusion draws on a very limited empirical material and, in general, it has been assumed, that through further empirical analysis, it would be possible to identify a field – or a number of fields – distinguishable from the CAP from which concepts and conceptions have been selected for translation. Hence, not only is the empirical analysis of possible processes of translation in the period after 1993 very limited, further empirical analysis may possibly also disclose that the spreading of concepts and conceptions holding, for instance, that organic farming at this point in time is faced by a number of problems, has progressed through processes of diffusion – as conceptualised from within the sociological institutionalism – rather than translation.

9. Related to the above, from certain quarters of the sociological but also the rational choice institutional perspective, it has been suggested that policy entrepreneurship is enabled by the ideational and institutional environment

within which it is exercised. It has also been suggested from a sociological institutional perspective that policy entrepreneurs may operate to link their particular 'pet solution' to 'the problem of the day' and that this is most likely to be successful during times of crisis. However, particularly with regard to the CAP it could be argued that: if ideas embedded in this field enjoy a high degree of institutionalisation then it may be expected that no single agent is able to enforce its particular ideals – even during times of crisis. Likewise, it could be argued that: if we are to conceptualise policy entrepreneurship – in relation to the study of ideas – in a way so that it is not confined to named individuals or collective agents, then we need to study the qualities of policy entrepreneurship.

The discursive institutional approach suggested has pursued the study of policy entrepreneurship as a position from which momentum is given to change and a threefold typology, distinguishing between the policy entrepreneurship exercised by translators, creators of forums for communication and carriers, has been suggested. Along these lines is has been possible to identify a number of agents, who at various points in time, and in various ways, have exercised policy entrepreneurship and, hence, given momentum to the articulation and institutionalisation of organic farming within the CAP. Additionally, it appeared that whereas various agents have exercised a particular type of policy entrepreneurship at particular points in time, policy entrepreneurship giving momentum to processes of articulation and institutionalisation of organic farming within the CAP in any given period is always exercised in concert among a number of agents. That said, in particular, the conclusions made in regard to the translators in the period from 1993 onwards should be seen against the background of the limited empirical analysis of possible processes of translation during this period. Moreover, the present study has been concerned with identifying various agents and various ways such agents have given momentum to change at various points in time. However, further theorisation is needed on the particularity of the discursive and institutional rules, which may enable policy entrepreneurship in the first place, and which might explain why some agents are able to exercise policy entrepreneurship while others, are not.

There are, of course, some obvious immediate explanations as to why the Commission, the Commission services and the Council are able to exercise policy entrepreneurship. Yet, it also appeared in the empirical analysis that a number of other actors, who are variously attributed with no or only limited formal decision-making powers, have given momentum to change at various point in time: most notably the EP and groupings within the EP. The theorisation of the discursive and institutional particularities enabling policy entrepreneurship would also have to be considered in the context of the conditions for and dynamics of institutional change suggested.

10. The final issue to be discussed regarding the limitations and virtues of the discursive institutional approach is related to the typology set up to distinguish between various ways that problems may appear within a discursive and institutional context. To be sure, the typology distinguishing between

theoretically, normatively, analytically and cognitively instituted problems does not imply that any particular type of problems should be more 'real' than others. That said it appeared in the empirical analysis that it was clearly possible to separate problems, which were based on, for instance, comprehensive statistical material, and certain widely accepted causal relationships such as problems of surplus productions, from problems which were, rather, based on certain wished/unwished for developments such as – at least in the mid-1980s – problems of environmental depletions. Although the typology in that sense has been helpful in distinguishing between various ways to generate problems, it is also clear that in reality and, for instance, in relation to particular policy proposals, the problems addressed may very well be more complex and, hence, difficult to characterise in a clear-cut way as either, for instance, cognitively or normatively instituted. It shall only be noted here that the study of institutional change along the lines of the discursive institutional approach pursued may possibly benefit from further theorisation on, for instance, the possible differences in the conditions conducive to change in the various types of problems as well as possible differences in their susceptibility to change. Arguably, cognitively instituted problems may be more susceptible to change than normatively instituted problems. For instance, the reality and significance of cognitively instituted problems must be expected to be evaluated in the context of available scientific evidence and statistical material, whereas normatively instituted problems are – when institutionalised – arguably less likely to be affected by such matters.

In conclusion, it could be argued that the current study possibly tends to challenge too many of the most common interpretations of what guides the development of the CAP. For instance, the common interpretation holding that change in the CAP is largely brought about by exogenous dynamics rather than by endogenous ones have been challenged. It has been proposed that perhaps the CAP – when it comes to the inflow of ideas – is not *as* sectorised a policy field as it is commonly considered to be. It has also been proposed that, for instance, the EP also has a role to play as an agent of change within the CAP, indicating that the Agriculture Council is, possibly, attributed too much significance as the defender of the status quo. In that sense, the current study probably gives rise to more questions, or perhaps hypotheses, about the conditions for, and dynamics of, institutional and ideational change than it answers.

References

Ackrill, Robert (2000) *The Common Agriculture Policy*. Sheffield Academic Press Ltd, UK.

Andersen, Niels Åkerstrøm (1995) *Selvskabt forvaltning: Forvaltningspolitikkens og central forvaltningens udvikling i Danmark 1900–1994*. Nyt fra Samfundsvidenskaberne, København, Denmark.

Andersen, Niels Åkerstrøm and Peter Kjær (1996) *Institutional Construction and Change: An Analytical Strategy of Institutional History*. COS-rapport no. 5/1996.

Andersen, Niels Åkerstrøm, Peter Kjær and Ove K. Pedersen (1996) On the Critique of Negotiated Economy. *Scandinavian Political Studies* 19, 2, pp. 167–177.

Banchoff, Thomas (2002) Institutions, Inertia and European Union Research Policy. *Journal of Common Market Studies* 40, 1, pp. 1–21.

Baumgartner, Frank R. and Barry D. Jones (1993) *Agendas and Instability in American Politics*. The University of Chicago Press, USA.

Bruckmeier, Karl and Wilking Ehlert (eds) (2002) *The Agri-environmental Policy of the European Union: The Implementation of the Agri-environmental Measures within the Common Agricultural Policy in France, Germany, and Portugal*. Europäischer Verlag der Wissenschaften, Germany.

Buller, Henry, Geoff A. Wilson and Andreas Höll (eds) (2000) *Agri-environmental Policy in the European Union*. Ashgate Publishing Limited, UK.

Campbell, John L. and Ove K. Pedersen (eds) (2001) *The Rise of Neoliberalism and Institutional Analysis*. Princeton University Press, New Jersey, USA.

Campbell, John L. and Ove K. Pedersen (2001a) The Rise of Neoliberalism and Institutional Analysis. In John L. Campbell and Ove K. Pedersen (eds) *The Rise of Neoliberalism and Institutional Analysis*. Princeton University Press, New Jersey, USA, pp. 1–23.

Campbell, John L. and Ove K. Pedersen (2001b) Conclusion. In John L. Campbell and Ove K. Pedersen (eds) *The Rise of Neoliberalism and Institutional Analysis*. Princeton University Press, New Jersey, USA, pp. 249–281.

Campbell, Hugh and Ruth Liepins (2001) Naming Organics: Understanding Organic standards in New Zealand as a Discursive Field. *Sociologia Ruralis* 41, 1, pp. 21–39.

Checkel, Jeffrey T. (1997) *Ideas and International Political Change: Soviet/Russian behaviour and the end of the cold war*. Yale University Press, New Haven, USA.

Checkel, Jeffrey T. (2001) Constructing European Institutions. In Gerald Schneider and Mark Aspinwall (eds) *The Rules of Integration:*

Institutionalist Approaches to the Study of Europe. Manchester University Press, Manchester, UK, pp. 19–39.

Cohen, Michael D., James G. March and Johan P. Olsen (1972) A Garbage Can Model of Organizational Choice. *Administrative Science Quarterly* 17, 1, pp. 1–25.

Coleman, William D. and Stefan Tangermann (1999) The 1992 CAP Reform, the Uruguay Round and the Commission: Conceptualising Linked Policy Games. *Journal of Common Market Studies* 37, 3, pp. 385–405.

Coleman, William D., Grace Skogstad and Michael M. Atkinson (1997) Paradigm Shifts and Policy Networks: Cumulative Change in Agriculture. In Wyn P. Grant and John T. S. Keeler (eds) (2000) *Agricultural Policy Volume I.* Edward Elgar Publishing Ltd, UK, pp. 120–148.

Coleman, William D., Wyn Grant and Timothy E. Josling (2004) *Agriculture in the New Global Economy.* Edward Elgar Publishing Ltd, UK.

Daugbjerg, Carsten (1998) *Policy Networks under Pressure: Pollution Control, Policy Reform and the Power of Farmers.* Ashgate Publishing Ltd, Aldershot, UK.

Daugbjerg, Carsten (1999) Reforming the CAP: Policy Networks and Broader Institutional Structures. *Journal of Common Market Studies* 37, 3, pp. 407–428.

Daugbjerg, Carsten (2003) Policy Feedback and Paradigm Shift in the EU Agricultural Policy: The Effects of the MacSharry Reform on Future Reform. *Journal of European Public Policy* 10, 3, pp. 421–437.

DiMaggio, Paul J. and Walter W. Powell (1991) The Iron Cage Revisited: Institutional Isomorphism and Collective Rationality in Organizational Fields. In Paul J. DiMaggio and Walter W. Powell (eds) *The New Institutionalism in Organizational Analysis.* The University of Chicago Press, USA, pp. 63–82.

Doron, Gideon and Itai Sened (2001) *Political Bargaining: Theory, Practice and Process.* Sage Publication Ltd, London, UK.

Downing, Keith (1994) The Compatibility of Behaviouralism, Rational Choice and New Institutionalism. *Journal of Theoretical Politics* 6, 1, pp. 105–117.

Downs, Anthony (1967) *Inside Bureaucracy.* Little, Brown and Company, Boston, USA.

Fennell, Rosemary (1985) Reconsideration of the Objectives of the Common Agriculture Policy. *Journal of Common Market Studies* 23, 3, pp. 257–276.

Fennell, Rosemary (1987) Reform of the CAP: Shadow or Substance?. *Journal of Common Market Studies* 26, 1, pp. 61–77.

Fennell, Rosemary (1997) *The Common Agriculture Policy: Continuity and Change.* Oxford University Press, Oxford, UK.

Flyvbjerg, Bent (2001) *Making Social Science Matter: Why Social Inquiry Fails and How it Can Succeed Again.* Cambridge University Press, Cambridge, UK.

Foster, Carolyn and Nicolas Lampkin (1999) European organic production statistics 1993–1996. *Organic Farming in Europe: Economics and Policy* 3, University of Hohenheim, Germany.

Foucault, Michel (1991) Politics and the Study of Discourse. In Graham Burchell, Colin Gordon and Peter Miller (eds) *The Foucault Effect: Studies in Governmentality*. Harvester Wheatsheaf, UK, pp. 53–72.

Fouilleux, E. (2004) Reforms and Multilateral Trade Negotiations: Another View on Discourse Efficiency. *West European Politics* 27, 2, pp. 235–255.

Gardner, Brian (1996) *European Agriculture: Policies, Production and Trade.* Routledge, London, UK.

Grant, Wyn (1993) Pressure Groups and the European Community: An Overview. In Sonia Mazey and Jeremy Richardson (eds) *Lobbying in the European Community*. Oxford University Press, Oxford, UK; pp. 27–46.

Grant, Wyn (1997) The Common Agriculture Policy, Macmillian, Basingstoke Press Ltd, UK.

Grant, Wyn, Dunkan Matthews and Peter Newell (2000) *The Effectiveness of European Union Environmental Policy*. Macmillian Press Ltd, Basingstoke, UK.

Greer, Alan (2005) *Agricultural Policy in Europe*. Manchester University Press, Manchester, UK.

Hajer, Maarten A. (1995) *The Politics of Environmental Discourse: Ecological Modernisation and the Policy Process*. Clarendon Press, Oxford, UK.

Hall, Peter A. (1989) The Politics of Keynesian Ideas. In Peter Hall A. (ed) *The Political Power of Economic Ideas: Keynesianism Across Nations.* Princeton University Press, New Jersey, USA, pp. 361–392.

Hall, Peter A. (1993) Policy Paradigms, Social Learning, and the State:The Case of Economic Policymaking in Britain. *Comparative Politics* 25, 3, pp. 275–296.

Hall, Peter A. and Rosemary C. R. Taylor (1996) Political Science and the Three New Institutionalisms. *Political Studies* XLIV, pp. 936–957.

Hay, Colin (2001) The 'Crisis' of Keynesianism and the Rise of Neoliberalism in Britain. In John L. Campbell and Ove K. Pedersen (eds) *The Rise of Neoliberalism and Institutional Analysis*. Princeton University Press, New Jersey, USA pp. 193–218.

Hennis, Marjoleine (2001) Europeanization and Globalization: The Missing Link. *Journal of Common Market Studies* 39, 5, pp. 829–850.

Hix, Simon (1999) *The Political System of the European Union*. Macmillan Press Ltd, Basingstoke, UK.

Jenkins-Smith Hank C. and Paul A. Sabatier (1993) The Dynamics of Policy-oriented Learning. In Paul A. Sabatier and Hank C. Jenkins-Smiths (eds) *Policy Change and Learning: An Advocacy Coalition Approach*. Westview Press Inc., Colorado, USA, pp. 41–56.

Jepperson, Ronald L. (1991) Institutions, Institutional Effects, and Institutionalism. In Paul J. DiMaggio and Walter W. Powell (eds) *The New*

Institutionalism in Organizational Analysis. The University of Chicago Press, USA, pp. 143–163.

Jones, Alun and Julian Clark (2001) *The Modalities of European Union Governance: New Institutionalist Explanations of Agri-environmental Policy.* Oxford University Press, Oxford, UK.

Kaltoft, Penille (2001) Organic Farming in the Late Modernity: At the Frontier of Modernity or Opposing Modernity?. *Sociologia Ruralis* 41, 1, pp. 146–158.

Kay, Adrian (1998) *The Reform of the Common Agricultural Policy.* CAB International, Wallingford, UK.

Kay, Adrian (2000) Towards a Theory of the Reform of the Common Agriculture Policy. *European Integration online Papers* 4, 9, http://eiop.or.at/eiop/texte/2000–009.htm.

Kay, A. (2003) Path dependency and the CAP. *Journal of European Public Policy* 10, 3, pp. 405–420.

Kingdon, John W. (1995) Agendas, Alternatives and Public Policies (2nd edition). Little, Brown and Company, Boston, USA.

Kjær, Peter and Ove K. Pedersen (2001) Translating Liberalization: Neoliberalism in the Danish negotiated economy. In John L. Campbell and Ove K. Pedersen (eds) *The Rise of Neoliberalism and Institutional Analysis.* Princeton University Press, New Jersey, USA pp. 219–248.

Kjær, Peter (1996) *The Constitutions of Enterprise: An Institutional History of Inter-firm Relations in Swedish Furniture Manufacturing.* Akademitryck AB, Stockholm University, Sweden.

Lampkin, Nicolas, Carolyn Foster, Susanne Padel and Peter Midmore (1999) The Policy and Regulatory Environment for Organic Farming in Europe. *Organic Farming in Europe: Economics and Policy* 1, University of Hohenheim, Germany.

Landau Alice (1998) Bargaining over Power and Policy: The CAP Reform and Agricultural Negotiations in the Uruguay Round. *International Negotiation* 3, 3, pp. 453–479.

Lenschow, Andrea (1998) The greening of the EU: The Common Agricultural Policy and Structural Funds. *Environment and Planning C: Government and Policy* 17, pp. 91–108.

Lenschow, Andrea and Anthony Zito (1998) Blurring or Shifting of Policy Frames?: Institutionalisation of the Economic–Environmental Policy Linkage in the European Community. *Governance* 11, 4, pp. 415–441.

Levitt, Barbara, and James G. March (1988) Organisational Learning. *Annual Review of Sociology* 14, pp. 319–340.

Lomborg, Bjørn (1998) *Verdens Sande Tilstand.* Centrum, Viby J., Denmark.

Lowe, Philip and Martin Whitby (1997) The CAP and the European Environment. In Christopher Ritson and David R. Harvey (eds) *The Common Agriculture Policy* (2nd edition). CAB International, Wallingford, UK pp. 285–304.

Lynggaard, Kennet (2001) The Farmer Within an Institutional Environment. Comparing Danish and Belgian Organic Farming. *Sociologia Ruralis* 41, 1, pp. 85–111.

March, James G. and Johan P. Olsen (1996) Institutional Perspectives on Political Institutions. *Governance: An International Journal of Policy and Administration* 9, 3, pp. 247–264.

March, James G. and Johan P. Olsen (1989) *Rediscovering Institutions: The organisational Basis of Politics*. Free Press, New York, USA.

McCormick, John (2001) *Environmental Policy in the European Union*. Palgrave, Basingstoke, UK.

Meadows, Donella, H., Dennis L. Meadows, Jørgen Randers, William W. Behrens III (1972) *TheLimits to Growth: A Report for the Club of Rome's Project on the Predicament of Mankind*. Leo Thorpe Limited, Wembley, UK.

Meyer, John W., John Boli and George M. Thomas (1994) Ontology and Rationalisation in the Western Cultural Account. In W. Richard Scott and John W. Meyer (eds) *Institutional Environments and Organizations: Structural Complexity and Individualism*. Sage Publications, USA, pp. 9–27.

Michelsen, Johannes (2001) Recent Development and Political Acceptance of Organic Farming in Europe. *Sociologia Ruralis* 41, 1, pp. 3–20.

Michelsen, Johannes, Kennet Lynggaard, Susanne Padel and Carolyn Foster (2001) Organic farming development and agricultural institutions in Europe: a study of six countries. *Organic Farming in Europe: Economics and Policy* 9. University of Hohenheim, Germany.

Minet, Paul (1962) *Full text of the Rome Treaty and An ABC of the Common Market*. Tileyard Press Ltd, London, UK.

Moyer, H. Wayne and Timothy E. Josling (1990) *Agricultural Policy Reform: Politics and Process in the EC and the USA*. Harvester Wheatsheaf, Hemel Hempstead, UK.

Nedergaard, Peter, Henning. O. Hansen and Preben Mikkelsen (1993) *EF's Landbrugspolitik og Danmark: Udviklingen frem til år 2000*. Handelshøjskolens Forlag, København, Denmark.

North, Douglass C. (1990) *Institutions, Institutional Change and Economic Performance*. Cambridge University Press, UK.

Nugent, Neill (1999) The Government and Politics of the European Union (4th edition).   Macmillan Press Ltd, Basingstoke, UK.

Olsen, Johan P. and B. Guy Peters (1996) Learning from Experience?. In Johan P. Olsen and B. Guy Peters (eds) *Lessons from Experience: Experiential Learning in Administrative Reforms in Eight Democracies*. Scandinavian University Press, Norway, pp. 1–35.

Olsen, Mancur (1971) *The Logic of Collective Action: Public Goods and the Theory of Groups*. Harvard University Press, USA.

Ostrom, Elinor (1990) *Governing the Commons: The Evolution of Institutions for Collective Action*. Cambridge University Press, UK.

Paarlberg, Robert (1997) Agricultural Policy Reform and the Uruguay Round: Synergistic Linkage in a Two-level Game?. *International Organization* 51, 3, pp. 413–444.

Padel, Susanne (2001) Conversion to Organic Farming: A Typical Example of the Diffusion of an Innovation?. *Sociologia Ruralis* 41, 1, pp. 40–61.

Pappi, Franz U. and Christian H. C. A. Henning (1999) The Organisation of Influence on the EC's Common Agricultural Policy: A Network Approach. *European Journal of Political Research* 36, 2, pp. 257–281.

Patterson, Lee Ann (1997) Agricultural Policy Reform in the European Community: A Three-level Game Analysis. *International Organization* 51, 1, pp. 135–165.

Pedersen, Ove K. (1988) Fra individ til aktør i struktur: Tilblivelse af en juridisk rolle. *Statsvetenskaplig Tidskrift* 3, Lund, Sweden, pp. 173–192.

Pedersen, Ove K. (1995) Problemets anatomi: eller problemet, der er et problem. *Tendens – Tidsskrift for Kultursociologi* 7, 1, pp. 10–21.

Peters, B. Guy (1996) Agenda-setting in the European Union. In Jeremy Richardson (ed) *European Union: Power and Policy-making*. Routledge, University of Essex, UK, pp. 61–71.

Peters, B. Guy (1999) *Institutional Theory in Political Science: The 'New Institutionalism'*. Pinter, London, UK.

Peterson, John and Elizabeth Bomberg (1999) *Decision-making in the European Union*. Macmillan Press Ltd, Basingstoke, UK.

Pierson, Paul (1998) The Path to European Integration: A historical-institutionalist Analysis. In Wayne Sanholtz and Alec Stone Sweet (eds) *European Integration and Supranational Governance*. Oxford University Press, Oxford, UK, pp. 27–58.

Pierson, Paul (2000) Increasing Returns, Path Dependence, and the Study of Politics. *American Political Science Review* 94, 2, pp. 251–267.

Pollack, Mark A. (1997) Delegation, Agency, and Agenda-setting in the European Community. *International Organization* 51, 1, pp. 99–134.

Pollack, Mark A. (2001) The Engines of Integration? Supranational Autonomy and Influence in the European Union. In Alec Stone Sweet, Wayne Sandholtz and Neil Fligstein (eds) *The Institutionalisation of Europe*. Oxford University Press, Oxford, UK, pp. 217–249.

Putnam, Robert D. (1988) Diplomacy and Domestic Politics: The Logic of Two-level Games. *International Organization* 42, 3, pp. 427–460.

Rochefort, David A. and Roger W. Cobb (eds) (1994) *The Politics of Problem Definition: Shaping the Policy Agenda*. University Press of Kansas, USA.

Rochefort, David A. and Roger W. Cobb (1994a) Problem Definition: An Emerging Perspective. In David A. Rochefort and Roger W. Cobb (eds) *The Politics of Problem Definition: Shaping the Policy Agenda*. University Press of Kansas, USA, pp. 1–31.

Roederer-Rynning, Christilla (2003) From 'Talking Shop' to 'Working Parliament'? The European Parliament and Agricultural Change. *Journal of Common Market Studies* 41, 1, pp. 113–135.

Roederer-Rynning Christilla (2003a) Informal Governance in the Common Agricultural Policy. In Thomas Christiansen and Simona Piattoni (eds) *Informal Governance in the European Union*. Edward Elgar Publishing Ltd, Cheltenham, UK, pp. 173–89.

Rosamond, Ben (2000) *Theories of the European Union*. Macmillan Press Ltd, Basingstoke, UK.

Rosamond, Ben (2002) Book Review and Notes. *Journal of Common Market Studies* 40, 2, pp. 363–364.

Sabatier, Paul A. and Hank C. Jenkins-Smiths (eds) (1993) *Policy Change and Learning: An Advocacy Coalition Approach*. Westview Press Inc, Colorado, USA.

Sabatier, Paul A. (1993) Policy Change over a Decade or More. In Paul A. Sabatier and Hank C. Jenkins-Smiths (eds) *Policy Change and Learning: An Advocacy Coalition Approach*. Westview Press Inc., Colorado, USA pp. 13–39.

Sbragia, Alberta M. (2000) Environmental Policy: Economic Constraints and External Pressure. In Helen Wallace and William Wallace (eds) *Policy-making in the European Union*. Oxford University Press, Oxford, UK, pp. 293–316.

Schneider, Gerald and Mark Aspinwall (eds) (2001) *The Rules of Integration: Institutionalist Approaches to the Study of Europe*. Manchester University Press, Manchester, UK.

Scott, W. Richard (1994) Institutions and Organisations: Towards a Theoretical Synthesis. In W. Richard Scott and John W. Meyer (eds) *Institutional Environments and Organizations: Structural Complexity and Individualism*. Sage Publications, USA, pp. 55–80.

Sheingate, Adam (2000) Agriculture Retrenchment Revisited: Issue Definition and Venue Change in the United States and European Union. *Governance* 13, 3, pp. 335–363.

Skogstad, Grace (1998) Ideas, Paradigms and Institutions: Agricultural Exceptionalism in the European Union and the United States. *Governance* 11, 4, pp. 463–490.

Skogstad, Grace (2001) The WTO and Food Safety Regulatory Policy Innovation in the European Union. *Journal of Common Market Studies* 39, 3, pp. 485–505.

Strang, David and John Meyer (1994) Institutional Conditions for Diffusion. In W. Richard Scott and John W. Meyer (eds) *Institutional Environments and Organizations: Structural Complexity and Individualism*. Sage Publications, USA, pp. 100–112.

Sverrisson, Árni (1999) Entrepreneurship and Brokerage: Translation, Networking and Novelty Construction in Ecological Modernisation.

Working Paper Series 66, Department of Sociology, University of Stockholm, Stockholm, www.sociology.su.se

Swinbank, Alan (1989) The Common Agriculture Policy and the Politics of Decision Making. *Journal of Common Market Studies* 27, 4, pp. 303–322.

Thelen, Kathleen and Sven Steinmo (1992) Historical Institutionalism in Comparative Politics. In Sven Steinmo, Kathleen Thelen and Frank Longstreth (eds) *Structuring Politics: Historical Institutionalism in Comparative Analysis.* Cambridge University Press, Cambridge, UK, pp. 1–32.

Weale, Albert, Geoffrey Pridham, Michelle Cini, Dimitrios Konstadakopulos, Martin Porter and Brendan Flynn (2000) *Environmental Governance in Europe: An Even Closer Ecological Union?.* Oxford University Press, Oxford, UK.

Wittrock Björn and Peter Wagner (1996) Social Science and the Building of the Early Welfare State. In Theda Skocpol and Dietrich Rueschemeyer (eds) *States, Social Knowledge, and the Origins of Modern Social Policies.* Princeton University Press, New Jersey, USA, pp. 90–113.

Documents

Agra Europe (116 articles from various issues of *Agra Europe* in the period from 19.10.1979 to 1.7.2005). A complete list of *Agra Europe* references is available on request.

Agri-food Partnership (2004) *Second Organic Action Plan for Wales 2005–2010.* Organic Centre Wales, Wales.

Biologica (1996) *Appél Groen Licht voor de Biologishe Landbouw. Platform Biologische Landbouw en Voeding, Oktober 1996.* Platform Biologische Landbouw en Voeding, Utrecht, The Netherlands.

Biologica (2000) *Rappél Groen Licht voor de Biologishe Landbouw Platform Biologische Landbouw en Voeding, August 2000.* Platform Biologische Landbouw en Voeding, Utrecht, The Netherlands.

CEPFAR (1996) *Proceedings of the European Seminar on: Organic Farming in the European Union, 6–8 June 1996.* CEPFAR, Vignola, Italy.

Commission (1981) Reflections on the Common Agriculture Policy, Commission communication to the Council presented on the 8 December 1980. *Bulletin of the European Communities, Supplement 6/80.* Office for Official Publications of the European Communities, Luxembourg.

Commission (1983) Common Agricultural Policy: Proposals of the Commission. *COM(83) 500 Final, 28.07.83.* Brussels.

Commission (1985) Communication from the Commission to the Council and Parliament: Perspectives for the Common Agricultural Policy. *COM(85) 333 Final, 15.07.85.* Office for Official Publications of the European Communities, Luxembourg.

Commission (1987) Proposal from the Commission on the prices for agriculture products and on related measures (1987/88) Volume 1, Explanatory memorandum. *COM(87) 1 final-Vol. I, 04.03.87*. Office for Official Publications of the European Communities, Luxembourg.

Commission (1988) Commission communication: Environment and agriculture. *COM(88) 338 final, 8.6.88*. Office for Official Publications of the European Communities, Luxembourg.

Commission (1989) Proposal for a Council Regulation (EEC) on organic production of agricultural products and indications referring thereto on agricultural products and foodstuffs. *COM(89) 552 final, 4.12.89*. Brussels.

Commission (1991) Communication of the Commission to the Council: The development and Future of the CAP Reflection Paper of the Commission. *COM(91) 100 final, 1.02.92*. Brussels.

Commission (1994) Organic Farming. *Green Europe 2/94*. Office for Official Publications of the EuropeanCommunities, Luxembourg.

Commission (1996), Proposal for a Council Regulation (EC) supplementing Regulation (CEE) No 2092/91 on organic production of agricultural products and indications referring thereto on agricultural products and foodstuffs to include livestock production. COM(96) 366 final. *Official Journal* C 293, 5.10.96, p. 0023.

Commission (1998) Proposals for Council Regulations (EC) concerning the reform of the common agriculture policy. *COM(1998) 158 final, 18.03.98*. Brussels.

Commission (1999) *Proceedings: Organic Farming in the European Union: Perspectives for the 21st Century, 27–28 May 1999*. Austrian Ministry for Agriculture and Forestry, Baden/Vienna, Austria.

Commission (1999a) CAP reform: A policy for the future. *Fact-sheet*. Brussel.

Commission (2001) *Økologisk Landbrug: Håndbog om Fælleskabets Lovgivning*. Kontoret for De Europæiske Fællesskabers Officielle Publikationer, Luxembourg.

Commission (2001a) Commission unveils '7-point plan' to tackle beef crisis. *Press release, 13.2.01*. Brussels.

Commission (2002) Analysis of the possibility of a European Action plan for organic food and farming. *Commission Staff Working Paper SEC (2002) 1368, 12.12.02*. Brussels.

Commission (2003). *COM (2003) 23 final, 21.1.03*. Brussels.

Commission (2003a) Report on the result of the online consultation: Action Plan for organic food and farming. http://europa.eu.int/comm/agriculture/qual/organic/plan/result_en.pdf (24.08.2005).

Commission (2004) European Hearing on Organic Food and Farming – Towards a European Action Plan, 22 January 2004. http://europa.eu.int/comm/agriculture/events/organic/index_en.htm (12.10.2005).

Commission (2004a) European Action Plan for Organic Food and Farming, Communication from the Commission to the Council and the European Parliament. *COM(2004) 415 final 10.06.2004*. Brussels,

Conference speech by Austrian Federal Minister for Agriculture and Forestry, Environment and Waterprotection W. Molterer. In Commission (1999) *Proceedings: Organic Farming in the European Union: Perspectives for the 21st Century, 27–28 May 1999*. Austrian Ministry for Agriculture and Forestry, Baden/Vienna, Austria.

Conference speech by Austrian Federal Minister for Agriculture and Forestry, Environment and Waterprotection W. Molterer. In Danish Ministry of Food (2001) *Proceedings: Organic Food and Farming: Towards Partnership and Action in Europe, 10–11 May 2001*.   Danish   Ministry   of   Food, Agriculture and Fisheries, Denmark.

Conference speech by Commissioner for Agriculture F. Fischler. In Commission (1999) *Proceedings: Organic   Farming   in   the   European   Union: Perspectives for the 21st Century, 27–28 May 1999*. Austrian Ministry for Agriculture and Forestry, Baden/Vienna, Austria.

Conference speech by Commissioner for the Environment R. Bjerregaard. In Commission (1999) *Proceedings: Organic Farming in the European Union: Perspectives for the 21st Century, 27–28 May 1999*. Austrian Ministry for Agriculture and Forestry, Baden/Vienna, Austria.

Conference speech by COPA representative in the Standing Committee for Organic farming T. Kaunisto.   In   Danish   Ministry   of   Food   (2001) *Proceedings: Organic Food and   Farming: Towards   Partnership   and Action in Europe, 10–11 May 2001*. Danish Ministry of Food, Agriculture and Fisheries, Denmark.

Conference speech by Danish Head of State P. N. Rasmussen. In Danish Ministry of Food (2001) *Proceedings: Organic Food and Farming: Towards Partnership and Action in Europe, 10–11 May 2001*. Danish Ministry of Food, Agriculture and Fisheries, Denmark.

Conference speech by Danish Minister for Food, Agriculture and Fisheries R. Bjerregaard. In Danish Ministry of Food (2001) *Proceedings: Organic Food and Farming: Towards Partnership and Action in Europe, 10–11 May 2001*. Danish Ministry of Food, Agriculture and Fisheries, Denmark.

Conference speech by Danish Ministry for Agriculture F. Matthiesen. In Commission (1999) *Proceedings: Organic Farming in the European Union: Perspectives for the 21st Century, 27–28 May 1999*. Austrian Ministry for Agriculture and Forestry, Baden/Vienna, Austria.

Conference speech by DG for Agriculture D. Givord. In Commission (1999) *Proceedings: Organic Farming in the European Union: Perspectives for the 21st Century, 27–28 May 1999.* Austrian   Ministry   for   Agriculture   and Forestry, Baden/Vienna, Austria.

Conference speech by DG for Agriculture P. Baillieux. In Commission (1999) *Proceedings: Organic Farming in the European Union: Perspectives for the*

21st Century, 27–28 May 1999. Austrian Ministry for Agriculture and Forestry, Baden/Vienna, Austria.

Conference speech by DG for Research Xabier Goenaga. In Danish Ministry of Food (2001) *Proceedings: Organic Food and Farming: Towards Partnership and Action in Europe, 10–11 May 2001.* Danish Ministry of Food, Agriculture and Fisheries, Denmark.

Conference speech by Director of Danish National Association of Organic Farming P. Holmbeck. In Danish Ministry of Food (2001) *Proceedings: Organic Food and Farming: Towards Partnership and Action in Europe, 10–11 May 2001.* Danish Ministry of Food, Agriculture and Fisheries, Denmark.

Conference speech by EEB G. Kuneman. In Danish Ministry of Food (2001) *Proceedings: Organic Food and Farming: Towards Partnership and Action in Europe, 10–11 May 2001.* Danish Ministry of Food, Agriculture and Fisheries, Denmark.

Conference speech by German Federal Minister for Consumer Protection, Nutrition and Agriculture R. Künast. In Danish Ministry of Food (2001) *Proceedings: Organic Food and Farming: Towards Partnership and Action in Europe, 10–11 May 2001.* Danish Ministry of Food, Agriculture and Fisheries, Denmark.

Conference speech by Greek Viceminister of Agriculture E. Argyris. In Danish Ministry of Food (2001) *Proceedings: Organic Food and Farming: Towards Partnership and Action in Europe, 10–11 May 2001.* Danish Ministry of Food, Agriculture and Fisheries, Denmark.

Conference speech by Head of Cabinet of DG for Agriculture C. Pirzio- Biroli. In Danish Ministry of Food (2001) *Proceedings: Organic Food and Farming: Towards Partnership and Action in Europe, 10–11 May 2001.* Danish Ministry of Food, Agriculture and Fisheries, Denmark.

Conference speech by President of IFOAM G. Rundgren. In Danish Ministry of Food (2001) *Proceedings: Organic Food and Farming: Towards Partnership and Action in Europe, 10–11 May 2001.* Danish Ministry of Food, Agriculture and Fisheries, Denmark.

Conference speech by Swedish Minister for Agriculture M. Wimberg. Danish Ministry of Food (2001) *Proceedings: Organic Food and Farming: Towards Partnership and Action in Europe, 10–11 May 2001.* Danish Ministry of Food, Agriculture and Fisheries, Denmark.

Conference speech by UK Junior Minister for Agriculture E. Morley. In Danish Ministry of Food (2001) *Proceedings: Organic Food and Farming: Towards Partnership and Action in Europe, 10–11 May 2001.* Danish Ministry of Food, Agriculture and Fisheries, Denmark.

Conference speech by Vicepresident of COPA P. Gæmelke. In Danish Ministry of Food (2001) *Proceedings: Organic Food and Farming: Towards Partnership and Action in Europe, 10–11 May 2001.* Danish Ministry of Food, Agriculture and Fisheries, Denmark.

Council (1973) Declaration of the Council of the European Communities and of the representatives of the Governments of the Member States meeting in the Council of 22 November 1973 on the programme of action of the European Communities on the environment. *Official Journal* C 112, 20.12.73.

Council (1977) Resolution of the Council of the European Community and of the Representatives of the Government of the Member   States   meeting within the Council of 17 May 1977 on the continuation and implementation of an European Community policy and action programme on the environment. *Official Journal* C 139, 13.06.77.

Council (1983) Council Decision 83/641/EEC of 12 December 1983 adopting joint research programmes and programmes for coordinating agricultural research. *Official Journal* L 358, 22.12.83.

Council (1985) Council Regulation (EEC) No 797/85 of 12 March 1985 on improving the efficiency of agricultural structures. *Official Journal* L 093, 30.03.85, pp.1–18.

Council (1991) Council Regulation (EEC) No. 2092/91 of 24 June 1991 on organic production of agriculture products and indications referring on agricultural products and foodstuffs. *Official Journal* L 198, 22.07.91, pp. 1–15.

Council (1992) Council Regulation No 2078/92 of 30 June 1992 on agricultural production methods compatible with the requirements of the protection of the environment and   the maintenance of the countryside. *Official Journal* L 215, 30.07.92, pp. 85–90.

Council (1999) Council Regulation (EC) No 1804/1999 of 19 July 1999 supplementing Regulation   (EEC) No 2092/91 on organic production of agricultural products and indications referring   thereto   on   agricultural products and foodstuffs to include livestock production. *Official Journal* L 222, 24.08.99, pp. 1–28.

Council (1999a) 2190th Council meeting: Agriculture, Luxembourg, 14 and 15 June 1999. *Press: 190 – Nr: 9000/99, 14.06.99.* Brussels.

Council (1999b) Council Regulation (EC) No 1257/1999 of 17 May 1999 on support for rural development from the European Agricultural Guidance and Guarantee Fund and   amending and repealing certain Regulations. *Official Journal* L 160,   22.06.99, pp. 80–102.

Council (1999c) 2218th Council meeting: Agriculture, Brussels, 15 November 1999. *Press: 345 –Nr: 12917/99, 15.11.99.* Brussels.

Council (2000) Organic farming: Follow-up discussion on the EU conference 'Organic Farming in the European Union: Perspective for the 21st Century' held in Vienna (Baden), 27–28 May 1999, *Annex submitted by Danish delegation in preparation of Agricultural meeting to be held 23–24 October 2000* (9.10.00, 12156/00). Brussels.

Council (2001) 2360th Council meeting: Agriculture, Luxembourg, 19 June 2001. *Draft minutes 10182/01, 23.07.2001.* Brussels.

Council (2002) 2476th Council meeting: Agriculture and Fisheries, Brussels, 16–20 December2002. *Press: 399 – Nr: 15636/02, 20.12.02*. Brussels.

Council (2003) 2516th Council meeting: Agriculture and Fisheries, Luxembourg, 11, 12, 17, 18, 19, 25, 26 June 2003. *Press: 164 – Nr: 10272/03*. Luxembourg.

Council (2004) 2592nd Council meeting: Agriculture and Fisheries, Luxembourg, 21 June 2004. *Press: 185 – Nr: 9999/04*. Luxembourg.

Danish Ministry for Food (2001) *Proceedings: Organic Food and Farming: Towards Partnership and Action in Europe, Copenhagen/Denmark 10–11 May 2001*. Danish Ministry of Food, Agriculture and Fisheries, Denmark.

DEFRA (2002) *Food and Farming: A sustainable future*. Policy Commission on the Future of Farming and Food, Department for Environment, Food and Rural Affairs, UK.

DEFRA (2002a) Action Plan to Develop Organic Food and Farming in England. Department for Environment, Food and Rural Affairs, UK.

DEFRA (2004) Action Plan to Develop Organic Food and Farming in England: two years on. Department for Environment, Food and Rural Affairs, UK.

DG for Agriculture (1996) The Community legislation concerning organic farming: Current state and future development. Conference Speech by A. Scharpe, DG for Agriculture. In *Proceedings of the European Seminar on: Organic Farming in the European Union, 6–8 June 1996*. CEPFAR, Vignola, Italy, pp. 3–16.

EEB (1978) *Conclusions drawn by the European Environmental Bureau from a Seminar on the Common Agriculture Policy, September 1978* (Reprint in pamphlet published by Soil Association, August 1979). Soil Association, Suffolk, UK.

Europa Parlamentet (1982) Beslutning om landdistriktudviklingens bidrag til genoprettelsen af den regionale ligevægt i Fællesskabet. *EF-tidende* C 66, 15.3.1982, pp. 21–25.

Europa Parlamentet (1986) Beslutning om landbrug og miljø. *EF-tidende* C 68, 24.3.1986, pp. 80–85.

European Parliament (1980) Written Question No. 1181/80 by Ernest Glinne to the Commission. *Official Journal* C 352, 31.12.80, pp. 5–6.

European Parliament (1981) On the contribution of rural development to the re-establishment of regional balances in the Community. *Report drawn up on behalf of the Committee on Regional Policy and Regional Planning*. Document 1-648 /81, 16.11.81.

European Parliament (1986) On agriculture and the environment. *Report on behalf of the Committee on the Environment, Public Health and Consumer Protection*. Document A 2-207/85, 03.02.86.

European Parliament (1987) Written Question No. 875/86 by Sylvester Barrett to the Commission. *Official Journal* C 31, 9.2.87, pp. 22–23.

European Parliament (1989) Written Question No. 409/88 by Vera Squarcialupi to the Commission. *Official Journal* C 132, 29.5.89, pp. 27–28.

European Parliament (1989a) Question No. 28 (H-474/89) by Scott-Hopkins to the Commission. *Official Journal* 3-384, 13.12.89, pp. 181–182.

European Parliament (1990) Written Question No. 296/89 by Lord O'Hagan to the Commission. *Official Journal* C 39, 19.2.90, pp. 6–7.

European Parliament (1990a) Written Question No. 540/90 by Mr José Happart to the Commission of the European Communities. *Official Journal* C 309, 10.12.90, p.9.

European Parliament (1990b) On the proposal from the Commission to the Council for a regulation on organic production of agricultural products and indications referring thereto on agricultural products and foodstuffs. *Report drawn up on behalf of the Committee on Agriculture, Fisheries and Rural Development.* Document A3-0311/90, 19.11.90.

European Parliament (1991) Written Question No. 1987/90 by Mr Gérard Monnier-Besombes to the Commission of the European Communities. *Official Journal* C 164, 24.06.91, p.9.

European Parliament (1991a) *Debates of the European Parliament.* No. 3-401, 19.2.91, pp. 76–78.

European Parliament (1997) On the proposal for a Council Regulation (EC) supplementing Regulation (EEC) No 2092/91 on organic production of agriculture products and indication referring thereto on agricultural products and foodstuffs to include livestock production. *Report drawn up on behalf Committee on Agriculture and Rural Development.* DOC_EN\RR\325\325619 PE 220.703/fin. 23.04.97.

European Parliament (1997a) Organic Production of agriculture products. *Debates of the European Parliament*, Sitting of 13.5.97.

European Parliament (1999) Organic Production. *Debates of the European Parliament*, Sitting of Wednesday 13.299.

European Voice (1.2.2001) Germany's urban cowgirl. *European Voice* 7, 5. The Economist Newspaper Limited, London, UK.

European Voice (22.11.2001) Künast challenges dyed-in-the-wool farming policies. *European Voice* 7, 43. The Economist Newspaper Limited, London, UK.

Financial Times (6.7.1996) Organic farm boost: Sally Smith looks at a market which may, at last, be about to take off. *Financial Times*, London, UK.

Financial Times (19.7.1996) Consumers in crisis go back to nature: Organic farming movement sees demand rise every month for meat and produce. *Financial Times*, London, UK.

Financial Times (20.7.1996) BSE scare boosts demand for organic food: Demand is so high 70% is imported from non-intensive farming suppliers overseas. *Financial Times* [US edition], London, UK.

Financial Times (19.1.2001) BSE persuades EU on eco-farming. *Financial Times*, London, UK.

Frankfurter Allgemeine (27.02.01) Thema der Woche: Die BSE- Krise. www.faz.com (March 2001).

House of Commons (1985) The UK Government agriculture development and advisory services, including lower input farming. *Agriculture Committee*, Session 1984–85. Her Majesty's Stationery Office, London.

House of Commons (2001) Organic Farming, Agriculture: Second Report. *Select Committee on Agriculture*, Session 2000–01. Her Majesty's Stationery Office, London.

IFOAM (1996) Organic Farming in the European Union, Conference Speech by B. Geier, General Secretary of IFOAM. In *Proceedings of the European Seminar on: Organic Farming in the European Union, 6–8 June 1996*. CEPFAR, Vignola, Italy, pp. 17–25.

IFOAM (2000) http://www.ifoam.org/standard/basics.html#6 (June 2003).

IFOAM (2003) IFOAM EU Regional Group, March 2003, http://www.ifoam.org/regional/eu_gp_intro0303.html (June 2003).

Jordbruksdepartementet (2004) Ekologisk Production. *Faktablad Augusti 2004*. Jordbruksdepartementet, Stockholm, Sweden.

Kommission (1968) Memorandum zur Reform der Landwirtschaft in der Europäischen Wirtschaftsgemeinschaft. *KOM(68) 1000, Teil A, 18.12.68*. Kommission der Europäischen Gemeinschaften, Brüssel.

Kommissionen (1985) En fremtid for EF's landbrug: Retningslinjer udarbejdet af Kommission på baggrund af konsultationerne I forbindelse med 'Grønbogen'. *KOM(85) 750 endelig udg., 18.12.85*. Kontoret for De europæiske Fælleskabers officielle Publikationer, Luxembourg.

Kommissionen (1988) Forslag til Rådets Forordning (EØF) om betingelser og bestemmelser for ydelse af støtte til omstilling af landbrugsproduktion. *Kom(88) 553 endelig udg., 21.10.88*, Brussels.

Living Earth (1990) Lady Eve Balfour: A tribute to an organic pioneer and the founder of the Soil Association. *The Journal of the Soil Association*, April–June 1990. Soil Association, Bristol, UK.

MAFF (1998) *MAFF: Sponsored Research into Organic Agriculture*. Ministry of Agriculture, Fisheries and Food, UK.

MAFF (1999) *Organic Farming*. Ministry of Agriculture, Fisheries and Food, UK.

Mansholt interview (1978) *The Common Agriculture Policy: Some New Thinking from Dr. Sicco Mansholt* (Reprint and translation from German, August 1979). Soil Association, Suffolk, UK.

Ministerie van Landbouw (2000) *An organic market to conquer (Een biologische markt te winnen)*. Ministerie van Landbouw, The Hague, The Netherlands.

Ministerie van Landbouw (2004) *Dutch Policy Document on Organic Agriculture 2005–2007*. Ministerie van Landbouw, The Hague, The Netherlands.

Organic Centre Wales (2003) http://www.organic.aber.ac.uk/stats.shtml (June 2003).

Scottish Executive (2003) *Organic Action Plan*. Scottish Executive, Scotland.

Soil Association (1979) Introduction. In *The Common Agriculture Policy: Some New Thinking from Dr. Sicco Mansholt*. Soil Association, Suffolk, UK.

Structure Directorate (1995) *Action Plan for the Aadvancement of Organic Food Production in Denmark*. Structure Directorate, Denmark.

Structure Directorate (1999) *Action Plan II: Developments in Organic Farming*. Structure Directorate, Denmark.

The Swedish Presidency (2003) http://eu2001.se/eu2001/main (June 2003).

Verbraucherministerium (2001) *Government policy statement on the new consumer protection and agricultural policies*. Speech by Renate Künast, Federal Minister on Consumer Protection, Food and Agriculture, 8.2.01. www.verbrauhermisterium.de (June 2003).

Welsh Organic Food Industry (1999) *Welsh Organic Food Sector: A Strategic Action Plan*. Welsh Organic Food Industry Working Group, Wales.

Index